D0321593

Macmillan Building and Surveying Series
Series Editor: IVOR H. SEELEY
Emeritus Professor, The Nottingham Trent University

Advanced Building Measurement, second edition Ivor H. Seeley
Advanced Valuation Diane Butler and David Richmond
An Introduction to Building Services Christopher A. Howard
Applied Valuation Diane Butler
Asset Valuation Michael Rayner
Building Economics, third edition Ivor H. Seeley
Building Maintenance, second edition Ivor H. Seeley
Building Procurement Alan E. Turner
Building Quantities Explained, fourth edition Ivor H. Seeley
Building Surveys, Reports and Dilapidations Ivor H. Seeley
Building Technology, fourth edition Ivor H. Seeley
Civil Engineering Contract Administration and Control Ivor H. Seeley
Civil Engineering Quantities, fifth edition Ivor H. Seeley
Civil Engineering Specification, second edition Ivor H. Seeley
Computers and Quantity Surveyors Adrian Smith
Contract Planning and Contractual Procedures, second edition B. Cooke
Contract Planning Case Studies B. Cooke
Design-Build Explained David E. L. Janssens
Development Site Evaluation N. P. Taylor
Environmental Science in Building, second edition R. McMullan
Housing Associations Helen Cope
Housing Management: Changing Practice edited by Christine Davies
Information Technology Applications in Commercial Property edited by Rosemary Feenan and Tim Dixon
Introduction to Valuation D. Richmond
Marketing and Property People Owen Bevan
Principles of Property Investment and Pricing W. D. Fraser
Property Valuation Techniques David Isaac and Terry Steley
Quality Assurance in Building Alan Griffith
Quantity Surveying Practice Ivor H. Seeley
Structural Detailing, second edition P. Newton
Urban Land Economics and Public Policy, fourth edition P. N. Balchin, J. L. Kieve and G. H. Bull
Urban Renewal – Theory and Practice Chris Couch
1980 JCT Standard Form of Building Contract, second edition R. F. Fellows

Building Quantities Explained

IVOR H. SEELEY

BSc(EstMan), MA, PhD, FRICS,
CEng, FICE, FCIOB, MIH
Chartered Quantity Surveyor
Emeritus Professor of The Nottingham Trent University

Fourth Edition

MACMILLAN

First published 1965 by
THE MACMILLAN PRESS LTD
Houndmills, Basingstoke, Hampshire RG21 2XS
and London
Companies and representatives
throughout the world

ISBN 0-333-48205-0 hardcover
ISBN 0-333-48206-9 paperback

A catalogue record for this book is available from the British Library.

Printed in Hong Kong

Reprinted 1966
SI edition 1969
Reprinted six times
Third edition 1979
Reprinted six times
Fourth edition 1988
Reprinted 1988, 1989, 1990, 1991 (twice), 1992, 1993

This book is dedicated to those numerous graduates, diplomates and students in quantity surveying of The Nottingham Trent University, with whom I have had the privilege and pleasure to be associated.

Est modus in rebus
which can be translated as
'there is measure/method in all things'
(motto of Royal Institution of Chartered Surveyors)

Contents

List of Figures

List of Examples

Preface to the Fourth Edition

The primary aim of this book continues to be to meet the needs of students studying the subject of building measurement in the earlier years of degree courses in quantity surveying and building, and those preparing for the first professional examinations of the relevant professional bodies and the Business and Technician Education Council. Although the needs of some of these students differ, it is believed that they will all benefit from the fundamental and yet practical approach that has been adopted.

This book covers the measurement of relatively simple building work in accordance with the principles laid down in the *Standard Method of Measurement of Building Works: Seventh Edition (SMM7)* supported by the associated *Code of Procedure*. The book contains a careful selection of worked examples of taking off, accompanied by comprehensive explanatory notes, covering all the basic work sections of the *Standard Method*. Its main aim is to simply explain and illustrate the method of measuring building work and to amplify and clarify the basic principles contained in the *Standard Method*, for the benefit of quantity surveying and building students. SMM7 contains a number of radical changes from previous practice and the book endeavours to highlight the main changes at appropriate points in the text and examples. Some clauses of the *Standard Method* are difficult to interpret and where more than one approach appears possible, the alternative procedures are described and examined.

In re-writing the book a positive aim, apart from complying with the prescribed requirements of SMM7 and thoroughly revising and updating the contents throughout, has been to produce a high standard of excellence in accuracy of dimensions, use of waste calculations, adequacy of descriptions, sequence of items and quality of presentation, which all aspiring students might seek to emulate. Furthermore, every effort has been made to incorporate and explain each step in the measurement process throughout the 25 worked examples, and to extend the usefulness of the book.

The book begins with a general introduction which traces the origins of the profession, the processes in use and an outline of the more recent developments, including project co-ordination documentation, followed by an examination of the general principles of taking off and the use of mensuration in building measurement. The bulk of the book is concerned with the practical application of the principles of the *Standard Method* together with the recognised taking off techniques to a wide range of building work. The final chapter examines and illustrates the various bill preparation processes ranging from abstracting and billing, including preambles and daywork, through to cut and shuffle, standard descriptions, the use of computers and other bill formats. Appendixes cover abbreviations, mensuration formulae, a metric conversion table and a manhole schedule, and there is a useful bibliography.

The unit symbol 'mm' has been omitted from all item descriptions, as well as drawing dimensions, since it now appears superfluous and its omission cannot give rise to confusion. However, they are rather surprisingly included in the billed examples contained in the *SMM7 Code of Procedure for Measurement of Building Works*.

The radical changes introduced by the Co-ordinating Committee for Project Information are examined and their possible implications and applications explored. It has to be emphasised that in practice a variety of approaches may be used and the author has attempted to adopt a standardised, logical approach following the format used in SMM7.

Nottingham,
Spring 1988 IVOR H. SEELEY

Acknowledgements

The author expresses his thanks to the Standing Joint Committee for the Standard Method of Measurement of Building Works for kind permission to quote from the *Standard Method of Measurement of Building Works: Seventh Edition (SMM7)*.

Christopher Willis FRICS FCIArb and Norman Wheatley FRICS, Chairman and Honorary Secretary respectively of the Standing Joint Committee, both gave friendly help and advice which was very much appreciated.

Ronald Sears MCIOB whom I have relied upon for many years past for his impressive artwork, once again prepared the handwritten dimensions and accompanying explanatory notes of outstanding quality on an extremely tight time schedule. His work does so much to bring the book to life as well as adding immensely to its value and quality.

Peter Murby and Malcolm Stewart of Macmillan Education Ltd have taken a keen personal interest in the production of the book. I am, as ever, greatly indebted to my wife for her continuing patience and understanding throughout the arduous re-writing period.

1 General Introduction

HISTORICAL BACKGROUND OF QUANTITY SURVEYING

The quantity surveying profession has largely developed over the last century, but has now grown to such an extent that it forms one of the largest single sections in the membership of the Royal Institution of Chartered Surveyors, particularly since the amalgamation with the former Institute of Quantity Surveyors in 1983. Quantity surveyors are employed in private practices, public offices and by contractors, and they undertake a diversity of work.

The earliest quantity surveying firm of which records are available is a Reading firm which was operating in 1785. There is little doubt that other firms were in existence at this time and a number of Scottish quantity surveyors met in 1802 and produced the first method of measurement. Up to the middle of the nineteenth century it was the practice to measure and value the building work after it had been completed and bills of quantities were not prepared.

The need for quantity surveyors became evident as building work increased in volume and building clients became dissatisfied with the method adopted for settling the cost of the work.

In the seventeenth century the architect was responsible for the erection of buildings, as well as their design, and he employed a number of master craftsmen who performed the work in each trade. Drawings were of a very sketchy nature and much of the work was ordered during the course of the job. On completion each master craftsman submitted an account for the materials used and labour employed on the work.

It later became the practice for many of the master craftsmen to engage 'surveyors' or 'measurers' to prepare these accounts. One of the major problems was to reconcile the amount of material listed on invoices with the quantity measured on completion of the work. Some of the craftsmen's surveyors made extravagant claims for waste of material in executing the work on the site and the architects also engaged surveyors to contest these claims.

General contractors became established during the period of the industrial revolution and they submitted inclusive estimates covering the work of all trades. Furthermore they engaged surveyors to prepare bills of quantities on which their estimates were based. As competitive tendering became more common the general contractors began to combine to appoint a single surveyor to prepare a bill of quantities, which all the contractors priced. In addition, the architect on behalf of the building owner usually appointed a second surveyor, who collaborated with the surveyor for the contractors in preparing the bill of quantities, which was used for tendering purposes.

In later years it became the practice to employ one surveyor only who prepared an accurate bill of quantities and measured any variations that arose during the progress of the project. This was the origin of the independent and impartial quantity surveyor as he operates today.

PURPOSES OF BILL OF QUANTITIES

Frequently, one of the principal functions of the client's quantity surveyor is the preparation of bills of quantities, although he does also perform a number of other functions which will be described later in this chapter. Consideration will now be given to the main purposes of a bill of quantities.

(1) First and foremost it enables all contractors tendering for a contract to price on exactly the same information with a minimum of effort.
(2) It limits the risk element borne by the contractor to the rates he enters in the bill.
(3) It prompts the client and the design team to finalise most project particulars before the bill can be prepared, ideally with full production drawings and project specification.
(4) It provides a basis for the valuation of variations which often occur during the progress of the work.
(5) It gives an itemised list of the component parts of the building, with a full description and the quantity of each part, and this may assist the successful contractor in ordering materials and assessing his labour requirements for the contract.
(6) After being priced it provides a good basis for a cost analysis, which subsequently will be of use on future contracts in cost planning work.
(7) If prepared in annotated form, it will help in the locational identification of the work.

It will be apparent that with the increasing size and complexity of building operations, it would be impossible for a contractor to price a medium to large sized project without a bill of quantities. For this reason it has been the practice for contractors to refrain from tendering in competition for all but the smallest contracts without bills of quantities being supplied. This approach does not apply to contracts for repairs or painting and decorating, where schedules of rates are usually more appropriate.

Furthermore, building projects even when they are concerned with the same type of building, usually vary considerably in detailed design, size, shape, materials used, site conditions and other aspects. For this reason a contractor could not readily give a price for building work, such as an office block, hospital or shop, based on the cost of a previous contract of similar type.

In the absence of a bill of quantities being prepared by the building owner, each contractor would have to prepare his own bill of quantities in the limited amount of time allowed for tendering. This places a heavy burden on each contractor and also involves him in additional cost which must be spread over the contracts in which he is successful. It could also result in higher cost to the client as contractors may feel compelled to increase their prices to cover the increased risks emanating from this approach.

PROCESSES USED IN QUANTITY SURVEYING WORK

The traditional method of preparation of a bill of quantities can conveniently be broken down into two main processes.

(1) 'Taking off', in which the dimensions are scaled or read from drawings and entered in a recognised form on specially ruled paper, called 'dimensions paper' (illustrated on page 16); and
(2) 'Working up' which comprises squaring the dimensions, as described in chapter 16, transferring the resultant lengths, areas and volumes to the abstract, where they are arranged in a convenient order for billing and reduced to the recognised units of measurement; and finally the billing operation, where the various items of work making up the complete project are listed in full, with the quantities involved in a suitable order under work section or elemental headings.

The most common approach is the group system (London method) whereby the work is measured in groups, each representing a particular section of the building without regard to the order in which the items will appear in the bill, as illustrated in the worked examples in this book. The alternative method is known as the trade by trade system (Northern method) when the taking off is carried out in trade order ready for direct billing, and thus eliminating the need for an abstract.

The term 'quantities' refers to the amounts of the different types of work fixed in position which collectively give the total requirements of the building contract.

These quantities are set down in a standard form on 'billing paper', as illustrated on page 229, which has been suitably ruled in columns, so that each item of work may be conveniently detailed with a description of the work, the quantity involved and a suitable reference. The billing paper also contains columns in which a contractor, tendering for a particular project enters the rates and prices for each item of work. These prices added together give the 'Contract Sum'.

The recognised units of measurement are detailed in the *Standard Method of Measurement of Building Works*, as listed in the bibliography at the end of the book. This document is extremely comprehensive and covers the majority of items of building work that are normally encountered. Many items are measured in metres and may be cubic, square or linear. Some items are enumerated and others, such as structural steelwork and steel reinforcing bars are measured by the tonne. The abbreviation SMM is used extensively throughout this book and refers to the *Standard Method of Measurement of Building Works: Seventh Edition (SMM7)*.

The bill of quantities thus sets down the various items of work in a logical sequence and recognised manner, so that they may be readily priced by contractors. A contractor will build up in detail a price for each item contained in the bill of quantities, allowing for the cost of the necessary labour, materials and plant, together with the probable wastage on materials and generally a percentage to cover establishment charges and profit. It is most important that each billed item should be so worded that there is no doubt at all in the mind of a contractor as to the nature and extent of the item which he is pricing. Contractors often tender in keen competition with one another and this calls for very skilful pricing to secure contracts.

The subject of estimating for building contracts is outside the scope of this book, but detailed information on this subject can be found in the books listed in the bibliography.

Civil engineering work is measured in accordance with the *Civil Engineering Standard Method of Measurement* (CESMM), and a useful textbook on this subject has been written by the same author, titled *Civil Engineering Quantities*.

Where a bill of quantities is prepared in connection with a building contract, it will almost invariably form a contract document to the exclusion of the specification, although much or all of the contents of a specification may be found in the preambles in a bill of quantities. The successful contractor is fully bound by the contents of all the contract documents when he signs the contract. The other contract documents on a normal building contract are the JCT Articles of Agreement and Conditions of Contract, Contract Drawings and Form of Tender.

Worked examples of more complicated types of building work including reinforced concrete, structural steelwork, mechanical and electrical services, and alteration work are contained in *Advanced Building Measurement*.

STANDARD METHOD OF MEASUREMENT OF BUILDING WORKS

The *Standard Method of Measurement of Building Works*, issued by the Royal Institution of Chartered Surveyors and the Building Employers Confederation forms the basis for the measurement of the bulk of building work.

The first edition was issued in 1922 with the expressed object of providing a uniform method of measurement based on the practice of the leading London quantity surveyors. Prior to the introduction of the first edition of the *Standard Method*, a large diversity of practice existed, varying with local custom and even with the idiosyncrasies of individual surveyors. This lack of uniformity afforded a just ground for complaint on the part of the contractors – that the estimator was often left in doubt as to the true meaning and intent of items in the bill of quantities which he was called upon to price, a condition which militated against scientific and accurate tendering.

It is interesting to note that this first nationally recognised *Standard Method of Measurement of Building Works* was prepared by representatives of the quantity surveyors and the building industry and that this Joint Committee also had consultations with representatives of certain trades. Building contractors have to price the bills of quantities and it is very desirable that they should be represented on the body which formulates the rules for measurement.

Further editions were issued in 1927, 1935, 1948, 1963, 1968, 1978 and the latest in 1988. All the references to the *Standard Method* in this book relate to the 1988 edition (SMM7).

The 1963 edition embraced a much greater number of changes with more far-reaching effect than had occurred in any previous revised edition. The rules previously covered by the supplements on Prestressed Concrete, Structural Steelwork, Terrazzo Work and Heating and Ventilating Engineer's Work were all incorporated and new sections were introduced covering Demolitions and Alterations, Piling, Underpinning and Fencing.

The sixth edition of the *Standard Method (SMM6)* was accompanied by a *Practice Manual* which, although non-mandatory, provided a guide to good practice, explaining and giving examples of how the rules in the SMM should be implemented.

Significant revisions were made to the Excavation and Earthworks, Concrete Works and Brickwork and Blockwork sections, while a new section of Woodwork replaced the previous Carpentry and Joinery sections, to reflect the change from a hand craft trade to a factory process. Revisions to other work sections were less extensive and were generally restricted to standardising phraseology and incorporating changes stemming from revisions of principle elsewhere, such as amendments to Drainage to correspond with Excavation and Earthworks. There was also more emphasis on drawn information than in earlier editions.

The Co-ordinating Committee for Project Information, sponsored by the Association of Consulting Engineers, Building Employers Confederation, Royal Institute of British Architects and Royal Institution of Chartered Surveyors produced a *Common Arrangement of Work Sections for Building Works* to be used in drafting specifications and bills of quantities. SMM7 has been structured on the basis so established as opposed to traditional works sections; hence the Standard Method of Measurement is now compatible with the other CCPI publications as described later in this chapter.

The other major change from previous editions of the SMM is that the measurement rules have been translated from prose into classification tables. This approach enables a quicker and more systematic use to be made of the measurement rules and readily lends itself to the use of standard phraseology and computerisation. The change does not, however, inhibit the use of traditional prose in the writing of bills of quantities where so desired. In addition to these two major changes, the rules have generally been simplified to produce shorter bills and the document has been updated to conform to modern practice.

The first section of the SMM incorporates *General Rules* which are of general applicability to all works sections and these are considered in detail in chapter 2.

Section A of the SMM is devoted to *Preliminaries and General Conditions* which incorporate general particulars of the project and the contract, including contractor's obligations, general arrangements relating to work by nominated sub-contractors, goods and materials from nominated suppliers and works by public bodies, and a list of general facilities and services which are included for convenience of pricing. The Preliminaries Bill is considered in greater detail in chapter 16.

CO-ORDINATED PROJECT INFORMATION

Research by the Building Research Establishment has shown that the biggest single cause of quality problems on building sites is unclear or missing project information. Another significant cause is unco-ordinated design, and on occasions much of the time of site management can be devoted to searching for missing information or reconciling inconsistencies in the data supplied.

The crux of the problem is that for most building projects the total package of information provided to the contractor for tendering and construction is produced in a variety of offices of different disciplines.

To overcome these weaknesses, the Co-ordinating Committee for Project Information (CCPI) was formed on the recommendation of the Project Information Group, sponsored by the four bodies listed previously (ACE, BEC, RIBA, and RICS). Its brief was to clarify, simplify and co-ordinate the national conventions used in the preparation of project documentation.

The following five documents have been published either by CCPI or by the separate sponsoring bodies.

(1) *Common Arrangement of Work Sections for Building Works.*
(2) *Project Specification — a code of procedure for building works.*
(3) *Production Drawings — a code of procedure for building works.*

(4) *Code of Procedure for Measurement of Building Works*.
(5) *SMM7*.

It is, however, unlikely that any single discipline office will require all these documents. For example, SMM7 conforms to the Common Arrangement and so quantity surveyors using the Standard Method will not of necessity require the latter document. Similarly, users of the National Building Specification (NBS) and the National Engineering Specification (NES) will not require the Common Arrangement.

Common Arrangement of work sections for building works

This document plays a major role in co-ordinating the arrangement of drawings, specifications and bills of quantities. It reflects the current pattern of sub-contracting and work organisation in building. To avoid problems of overlap between similar or related work sections, each section contains a comprehensive list of what is included in the section and what is excluded, stating the appropriate section of the excluded items.

SMM7 uses the same work sections and this will eliminate any inconsistencies between specifications and bills of quantities, where the quantity surveyor structures the bill on SMM7.

The *Common Arrangement* has a hierarchical arrangement in three levels, for example:

Level 1: D Groundwork
Level 2: D3 Piling
Level 3: D30 Cast-in-place concrete piling (work section).

It lists 24 level 1 group headings and about 300 work sections, roughly equally divided between building fabric and services. However, no single project will encompass more than a relatively small number of them. Only levels 1 and 3 will normally be used in specifications and bills of quantities, while level 2 allows for the insertion of new works sections if required later, without recourse to extensive renumbering.

Common Arrangement describes how a work section is a dual concept, involving the resources being used and also the parts of the work being constructed, including their essential functions. The category is usually influenced and characterised by both input and output. For example, an input of brick or block could have an output of walling, while an input of mastic asphalt could have an output of tanking.

Section numbers are kept short for ease of reference. The widespread use of cross-references to the specification should encourage designers to be more consistent in the amount of description which they provide on drawings.

Project Specification — a code of procedure for building works

This code draws a distinction between specification information and the project specification. Information may be provided on drawings, in bills of quantities or schedules, but the project specifications should be the first point of reference when details of the type and quality of materials and work are required. Hence drawings and bills of quantities should identify kinds of work but not specify them. Instead, simple cross-references should ideally be made to the specification as, for example, Ledkore damp-proof course F30.2.

The project specification should be prepared by the designer and the use of a standard library of specification clauses will make the task easier. Specifications should be arranged on the basis of the *Common Arrangement*. Both the NBS library of clauses and the NES for services installations are so arranged.

The code provides extensive check lists for the specification of each work section to ensure that project specifications are complete, and it also gives advice on specification preparation by reference to British Standards or other published documents or by description.

Production Drawings — a code for building works

This code deals with the management of the preparation, co-ordination and issue of sets of drawings, and with the programming of the design and communication operation, and thus complements BS 1192: 1984 (Construction Drawing Practice).

The following criteria should preferably be adopted in the preparation of drawings:

(1) Use of common terminology. If the content of a drawing coincides with a Common Arrangement work section, then the Common Arrangement title should be used on that drawing.
(2) Annotate drawings by cross reference to specification clause numbers, for example, concrete mix A, E.10.4 and lead flashing, H/1.

SMM7 and Code of Procedure for Measurement of Building Works

The arrangement of SMM7 is based on the *Common Arrangement* and the rules of measurement for each work section are in the same sequence.

If descriptions in the bills of quantities are cross-referenced to clause numbers in the specification, for example concrete mix A, E10.4, as for the drawings, then the co-ordination of drawings, specifications and bills of quantities will be improved, and the risk of inconsistent information will be reduced. If required, the specification can be incorporated into the bill of quantities as preambles.

The code of procedure explains and enlarges upon the SMM as necessary and gives guidance on the arrangement of bills of quantities. It should be emphasised that the code is for guidance only and does not have the mandatory status of SMM7.

Standard Classifications

The international SfB system is scheduled to be revised in 1990, and it is possible that the level 1 headings from the *Common Arrangement* may be incorporated in the revised version. The incorporation of the *Common Arrangement* into CI/SfB, the United Kingdom version of SfB, at its next major revision is under active consideration.

There are also indications that the structure of British Standard specifications may be based on the *Common Arrangement*, so further extending the scope of co-ordinated project information.

Co-ordinated Project Information in Use

Since the *Common Arrangement* is based on natural groupings of work within the building industry, it is likely to provide benefits in the management of the construction stages. Not only will it be much easier to find the required information, but it will also be structured by the *Common Arrangement* in a manner which conforms to normal sub-contracting and specialist contracting practice. Thus in obtaining estimates from sub-contractors, it will be a straightforward task to assemble the correct set of drawings, specification clauses and bill items. In management contracting, the *Common Arrangment* is likely to provide a convenient means of identifying separate work packages. Similarly, construction programmes based on the *Common Arrangement* will provide direct links to other project information, thus bringing together quantity, cost and time data into an integrated information package.

Further standardisation is being introduced through the publication of *SMM7 Library of Standard Descriptions*, jointly sponsored by the Property Services Agency (DOE), Royal Institution of Chartered Surveyors and Building Employers Confederation. However, quantity surveyors are not obliged to refer to any of the Co-ordinated Project Information (CPI) documents apart from SMM7 when producing bills of quantities. Hence, in practice, a variety of approaches can be adopted when framing billed descriptions including cross-references to project specification clauses, use of the *SMM7 Library of Standard Descriptions*, individual descriptions built up from SMM7 by quantity surveyors as implemented in the worked examples in this textbook, traditional prose given as another acceptable option in the preface to SMM7, or the use of *Shorter Bills of Quantities: The Concise Standard Phraseology and Library of Descriptions*.

The latter publication in two extensive volumes, prepared by two eminent quantity surveying practices, seeks to provide guidelines for the measurement of projects of simpler design and/or construction, where less detailed measurement will suffice, and offers further limited additional simplification of the measurement rules compared with SMM7. It contains very clear classification tables with up to six levels of description with numerous specification examples and an extensive library of standard descriptions. It does not however have the official approval of the SMM7 sponsoring bodies but this is unlikely to preclude its use by some quantity surveying practices.

OTHER FUNCTIONS OF THE QUANTITY SURVEYOR

The *client's quantity surveyor* performs a variety of functions as now listed, and the underlying theme of a quantity surveyor's work is one of cost control rather than the preparation of bills of quantities and settlement of final accounts, whether he be engaged in private practice or in the public sector.

(1) Preparing approximate estimates of cost in the very early stages of the formulation of a building project, and giving advice on alternative materials and types of construction and the financial aspects of contracts.
(2) Cost planning during the design stage of a project to ensure that the client obtains the best possible value for his money, preferably having regard to total costs using life cycle costing techniques, that the costs are distributed in the most realistic way throughout the various sections or elements of the building and that the tender figure is kept within the client's budget, as described in *Building Economics*.
(3) Examining tenders and priced bills of quantities and reporting his findings.
(4) Negotiating rates with contractors on negotiated contracts and dealing with cost reimbursement contracts, design and build, management and other forms of contract.
(5) Valuing work in progress and making recommendations as to payments to be made to the contractor, including advising on the financial effect of variations.
(6) Preparing the final account on completion of the contract works.
(7) Advising on the financial and contractual aspects of contactors' claims.
(8) Giving cost advice and information at all stages of the contract and preparing cost analyses.

The *contractor's quantity surveyor* performs a rather different range of functions, and these are now described, since there can be few of the large or medium size contracting firms who do not employ quantity surveyors. Usually, the contractor's organisation will include a quantity surveying department controlled by a qualified quantity surveyor who is normally a senior executive and may have director status.

The duties of the contractor's quantity surveyor will vary according to the size of the company employing him; tending to be very wide in scope with the smaller companies, but rather more specialised with the larger firms.

In the smaller company, his activities will be of a general nature and include preparing bills of quantities for small contracts; agreeing measurements with the client's quantity surveyor; collecting information about the cost of various operations from which the contractor can prepare future estimates; preparing precise details of the materials required for the projects in hand; compiling target figures so that the operatives can be awarded production bonuses; preparing interim costings so that the financial position of the project can be ascertained as the work proceeds and appropriate action taken where necessary; planning contracts and preparing progress charts in conjunction with the general foremen; making application to the architect for variation orders if drawings or site instructions vary the work; agreeing sub-contractors' accounts; comparing the costs of alternative methods of carrying out various operations so that the most economical procedure can be adopted.

In larger companies, the contractor's quantity surveyor will not usually cover such a wide range of activities since different departments handle specific activities. During his training period, the trainee quantity surveyor will probably spend some time in each of the departments.

Readers requiring more detailed information about quantity surveying functions are referred to *Quantity Surveying Practice*.

MODERN QUANTITY SURVEYING TECHNIQUES

A number of developments have taken place in the method of preparation and form of bills of quantities in recent years.

Methods of Preparation of Bills of Quantities

The traditional method of taking off, abstracting and billing is both lengthy and tedious in the extreme. Alternative systems have accordingly been introduced with a view to speeding up the process, lowering cost and reducing the requirements for working up staff, who are in short supply, as the large workforce of quantity surveying technicians envisaged in the early nineteen seventies has not materialised, mainly because of the lack of adequate status.

The two principal improved methods of bill preparation are cut and shuffle and the use of computers; both these methods will be considered in some detail in chapter 16. As long ago as 1962, a working party of the Royal Institution of Chartered Surveyors was of the opinion that the quantity surveying profession could and ought to take advantage of mechanical and other aids which were available or could be developed, to economise in the use of labour and accelerate the production of bills of quantities, and there have been substantial developments since that time.

The introduction of modern systems for preparing bills of quantities has led to the transfer of staff in quantity surveyors' offices from working up to other types of work. Different methods of training quantity surveyors have been developed and increasing numbers of students are attending full-time and thick sandwich quantity surveying degree and diploma courses at universities, polytechnics and other colleges. There has also been a progressive increase in the number of part-time degree and diploma courses and significant improvements in the scope and form of distance learning, and continuing professional development facilities.

Greater standardisation in the presentation of bills of quantities, both as regards order of billing and method of presenting items, is now considered vitally important and this, coupled with more uniformity in contractors' methods of estimating, costing and programming, enables the fullest use to be made of computerised systems.

Formats of Bills of Quantities

In the post-war years a number of different formats of bills of quantities have been used, in an endeavour to produce a bill of quantities which would be of greater value to the contractor than the normal work section order bill of quantities.

One development was the use of elemental bills in which the items were arranged, not in work section order under main headings of the separate work sections, but were grouped according to their position in the building (elements). Each element comprised an integral part of the building such as external walls, roofs or floors, each of which performed a certain design function. Within each element, the items could be billed in work section order or grouped in building sequence.

It was intended that this format of bill would assist in more precise tendering, help on the project by locating the work in the bill and make the operation of cost planning and cost analysis techniques much easier.

In practice it was found that contractors did not favour the use of elemental bills, principally on the grounds that possible advantages on the site were outweighed by disadvantages at the tendering stage, and this was a major item when only about 5 per cent of a contractor's tenders were likely to prove successful. The major disadvantage to the estimator was that he had to look through many of the elements, to collect together all the items in a given work section, before he could assess the total quantities of materials or type of plant required; and similar difficulties arose with work which was to be sub-let.

Another type of bill which could be presented either as a trade bill or with elements as the main sub-division was described as a sectionalised trades bill, and this met some of the contractors' objections to elemental bills.

Yet another bill format, known as the operational bill was developed by the Building Research Establishment. In this form of bill the description of the billed work followed the actual building process, with materials shown separately from labour, all described in terms of the operations necessary for the construction of the building. An operation for the purpose of the operational bill was the work performed by a man or gang between definite breaks in the work pattern, such as bricklaying from damp-proof course to first floor joists in housing work.

In practice, the disadvantages of operational bills outweighed their advantages and they have been little used. Other bill formats included activity bills and annotated bills and they are described in some detail, along with other developments, in chapter 16.

As described earlier in the chapter, there have also been significant developments in the standardisation of descriptions of bill items; started by Fletcher and Moore in their *Standard Phraseology for Bills of Quantities*. The latest development has been the production of the *SMM7 Library of Standard Descriptions*.

2 General Principles of Taking Off

GENERAL RULES

Basic Principles

Some of the general principles to be followed in taking off building quantities are detailed in the General Rules in the first section of the *Standard Method of Measurement of Building Works*, of which the following statements are particularly important.

> This Standard Method of Measurement provides a uniform basis for measuring building works and embodies the essentials of good practice. Bills of quantities shall fully describe and accurately represent the quantity and quality of the works to be carried out. More detailed information than is required by these rules shall be given where necessary in order to define the precise nature and extent of the required work (SMM General Rules 1.1).
>
> Rules of measurement adopted for work not covered by these rules shall be stated in a bill of quantities. Such rules shall, as far as possible, conform with those given in this document for similar work (SMM General Rules 11.1).

The billed descriptions are to be reasonably comprehensive and sufficient to enable the estimator fully to understand what is required and to give a realistic price. All quantities must be as accurate as the information available permits, as inaccurate bills cause major problems.

It is most important that all work whose extent cannot be determined with a reasonable degree of accuracy should be described as approximate quantities, and items of this kind should be kept separate from those which contain accurate quantities (SMM General Rules 10.1). In this way the contractor is made aware of the uncertain nature of the quantity entered and that it will be subject to re-measurement on completion and valuation at billed rates. This can apply to any work where the architect is unable to give full details at the time of measurement.

In SMM General Rules 3.1–3, it is emphasised that measurements are to relate to work net as fixed in position, except where otherwise described in a measurement rule applicable to the work. Measurements are to be taken to the nearest 10 mm (5 mm and over shall be regarded as 10 mm and less than 5 mm shall be disregarded). Lengths are entered in the dimension column in metres to two places of decimals. When billing in metres the quantity is billed to the nearest whole unit, but where the unit of billing is the tonne, quantities shall be billed to two places of decimals.

Where a measurement rule provides that the area or volume comprising a void is not deducted from the area or volume of the surrounding material, for example $\leq 1.00\ m^2$ for roof coverings, this shall refer only to openings or wants which are within the boundaries of the measured areas. Openings or wants which are at the boundaries of measured areas shall always be the subject of deduction irrespective of size (SMM General Rules 3.4).

Billed items are generally deemed to include, that is, without the need for specific mention: labour, materials and plant goods, including all associated costs; waste of materials, square cutting; establishment and overhead charges and profit (SMM General Rules 4.6).

Each work section of a bill of quantities shall begin with a description stating the nature and location of the work, unless evident from the drawn or other information required to be provided by the SMM rules (SMM General Rules 4.5).

Four categories of drawing are listed in SMM General Rules 5.1–4 (location drawings, component drawings, dimensioned diagrams and schedules). The student will be particularly concerned with the third category — dimensioned diagrams — to show the shape and dimensions of the work covered by an item, and they may be used in a bill of quantities in place of a dimensioned description, but not to replace an item otherwise required to be measured. These drawings will be considered further when dealing with specific examples later in the book.

Tabulated Rules

The rules prescribed in SMM7 are set out in the form of tables and these comprise classification tables and supplementary rules. Horizontal lines divide the classification table and supplementary rules into zones to which different rules apply. Where broken horizontal lines appear within a classification table, the rules entered above and below these lines may be used as alternatives (SMM General Rules 2.1–3). As, for example, metal sheet flashings, aprons, cappings and the like which may be measured either with a dimensioned description or with a dimensioned diagram (SMM H70.10–18.1–2). Within the supplementary rules everything above the horizontal line, which is immediately below the classification table heading, is applicable throughout that table (SMM General Rules 2.8).

The left-hand column of a classification table lists descriptive features commonly encountered in building works, followed by the relevant unit of measurement. The next or second column lists sub-groups into which each main group shall be divided and the third column provides for further subdivision, although these lists are not intended to be exhaustive. Each item description shall identify the work relating to one descriptive feature drawn from each of the first three columns in the classification table, and as many of the features in the fourth or last column as are appropriate. Where the abbreviation (nr) is given in the classification table, that quantity shall be stated in the item description (SMM General Rules 2.4–7).

The supplementary rules form an extension of the classification tables and are sub-divided into the following four columns:

(1) measurement rules prescribe when and how work shall be measured;
(2) definition rules define the extent and limits of the work contained in the rules and subsequently used in the preparation of bills of quantities;
(3) coverage rules draw attention to incidental work which is deemed to be included in appropriate items in the bill of quantities to the extent that such work is included in the project, and where coverage rules include materials they shall be mentioned in item descriptions;
(4) supplementary information contains rules covering any additional information that is required (SMM General Rules 2.9–12).

Cross references within the classification tables encompass the numbers from the four columns, such as D20.2.6.2.0: excavating trenches; width > 0.30 m; maximum depth ≤ 1.00 m. The digit 0 indicates that there are no entries in the column in which it appears, while an asterisk represents all entries to the column in which it occurs (SMM General Rules 12.2–4).

DIMENSIONS PAPER

The normal ruling of dimensions paper on which the dimensions, as scaled or taken direct from drawings, are entered, is now indicated. This ruling conforms to the requirements of BS 3327: 1970 *Stationery for Quantity Surveying*, showing the face side of the sheet with a binding margin on the left-hand side.

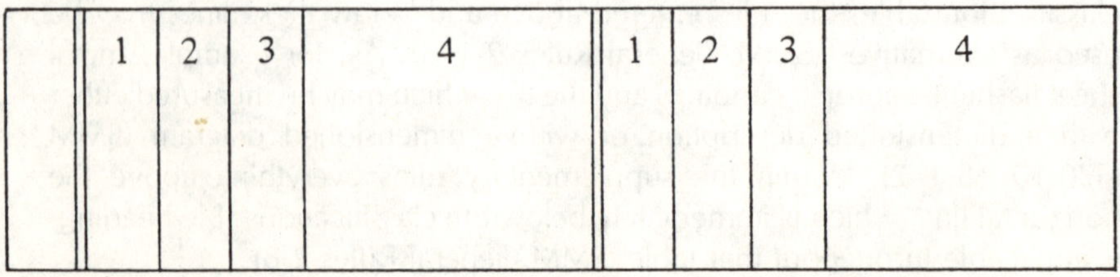

	1	2	3	4	1	2	3	4

Each dimension sheet is split into two identically ruled parts, each consisting of four columns. The purpose of each column will now be indicated for the benefit of those readers who are unfamiliar with the use of this type of paper.

Column 1 is called the 'timesing column' in which multiplying figures are entered when there is more than one of the particular item being measured.

Column 2 is called the 'dimension column' in which the actual dimensions, as scaled or taken direct from the drawings, are entered. There may be one, two or three lines of dimensions in an item depending on whether it is linear, square or cubic.

Column 3 is called the 'squaring column' in which the length, area or volume obtained by multiplying together the figures in columns 1 and 2 is recorded, ready for transfer to the abstract or bill.

Column 4 is called the 'description column' in which the written description of each item is entered. The right-hand side of this wider column is frequently used to accommodate preliminary calculations and other basic information needed in building up the dimensions and references to the location of the work, and is referred to as 'waste'.

In the worked examples that follow in succeeding chapters the reader will notice that one set of columns only is used on each dimension sheet, with the remainder used for explanatory notes, but in practice both sets of columns will be used for taking off.

Dimensions paper is almost universally of international paper size A4 (210 mm × 297 mm).

An alternative approach is to use some form of cut and shuffle sheets as described and illustrated in chapter 16.

Entering Dimensions

Spacing of Items

It is essential that ample space is left between all items on the dimension sheets so that it is possible to follow the dimensions easily and to enable any items, which may have been omitted when the dimensions were first taken off, to be subsequently inserted without cramping the dimensions unduly. The cramping of dimensions is a common failing among examination candidates and does cause loss of marks. The items contained in the worked examples in this book are often closer than is ideal, solely to conserve space and keep down the price of this student textbook!

Waste

The use of the right-hand side of the description column for preliminary calculations, build up of lengths, explanatory notes, location of measured work, and the like should not be overlooked. All steps that have been taken in arriving at dimensions, no matter how elementary or trivial they may appear, should be entered in the waste section of the description column. Following this procedure will do much to prevent doubts and misunderstandings concerning dimensions arising at some future date. It also enables all calculations for dimensions to be checked.

Order of Dimensions

A constant order of entering dimensions should be maintained throughout in accordance with SMM General Rules 4.1, that is, (i) length, (ii) width or breadth and (iii) height or depth, even although the SMM requirement strictly relates only to dimensions in descriptions. In this way there can be no doubt as to the shape of the item being measured. When measuring a cubic item of concrete 3.50 m long, 2.50 m wide and 0.50 m deep, the entry in the dimension column would be

	3.50 2.50 0.50		*In situ* conc., class A, isoltd. fdns.

It will be noted that dimensions are usually recorded in metres to two places of decimals with a dot between the metres and fractions and a line drawn across the dimension column under each set of figures.

The unit symbol mm has been omitted from all descriptions since it is considered superfluous and its omission should not cause any confusion, despite its inclusion in the examples in the *Code of Procedure for Measurement of Building Works*.

Timesing

If there were twelve such items, then this dimension would be multiplied by twelve in the timesing column, as in the following example

12/	3.50 2.50 0.50		*In situ* conc., class A, isoltd. fdns.

If it was subsequently found that four more foundation bases of the same dimensions were to be provided, then a further four could be added in the timesing column by the process known as 'dotting on', as indicated in the next example.

12/ 4˙	3.50 2.50 0.50	*In situ* conc., class A, isoltd. fdns.	

Abbreviations and Symbols

Many of the words entered in the description column are abbreviated in order to save space and time spent in entering the items by highly skilled technical staff. Many abbreviations have become almost standard and are of general application; for this reason a list of the more common abbreviations is given in appendix I at the end of this book. A considerable number of abbreviations are obtained by merely shortening the particular words, such as the use of 'conc.' in place of concrete and 'rad.' for radius. With some measurement techniques, such as cut and shuffle, it may be considered desirable to avoid the use of abbreviations where bill descriptions are to be typed direct from the initial measurement or dimension sheets or slips. Abbreviations save time in examinations.

An extensive list of symbols is given in SMM General Rules 12.1 and include m (metre), m^2 (sq.m), m^3 (cu.m), mm (millimetre), nr (number), t (tonne), h (hour), $>$ (exceeding), $\leq$ (not exceeding), $<$ (less than), % (percentage) and – (hyphen; often used to denote range of dimensions).

Grouping of Dimensions

Where more than one set of dimensions relate to the same description, the dimensions should be suitably bracketed so that this shall be made clear. The following example illustrates this point.

			Mhs.
	18.60	Pipes in trs.	(1–2
	25.00	100 BS clay	(2–3
	42.60	to BS 65,	(3–4
	36.00	Pt. 2, jtd.	(4–5
		w. flex. mech. jts.	

Note also the location particulars entered in waste which readily identify the location of each length of drain.

Where the same dimensions apply to more than one item, the best procedure is to separate each of the descriptions by an '&' sign and to bracket the descriptions, as illustrated in the following example.

	35.00 0.75 0.90		Exc. tr., width > 0.30 m, max. depth ≤ 1.00 m. & Fillg. to excavns., av. thickness > 0.25 m, arisg. from excavns.

Deductions

After measuring an item it is sometimes necessary to deduct for voids or openings in the main area or volume. This is normally performed by following the main item by a deduction item as shown in the following example.

	21.30 20.30		Exc. topsoil for preservn. av. 150 dp.
	7.60 4.60 5.00 5.00		Ddt. ditto.

(Note the underlining of the word deduct)

Measurement of Irregular Figures

It is sometimes necessary to measure the areas of triangles and circles, the circumferences of circles, and the volumes of cylinders, and the like, and the usual method of entering the dimensions is illustrated in the following examples.

½/	4.00 3.00		Area of triangle with a base of 4 m and a height of 3 m. (area = base × ½ height)
22/7/	2.00 2.00		Area of circle, 2 m radius. (area = πr^2)
½/2/22/7/	2.00		Circumference of semi-circle, 2 m radius. (circumference of whole circle = $2\pi r$)
22/7/	0.50 0.50 3.00		Volume of cylinder, 1 m diameter and 3 m high. (area of circle × height of cylinder)

Alterations to Dimensions

It is sometimes necessary to substitute amended dimensions in place of those which have already been entered on the dimensions paper. The student is advised never to alter the original figures, because, apart from looking most untidy, it is often extremely difficult to decipher the correct figures. If it is necessary to amend figures one procedure is to cross out the original figures and neatly write the new figures above them, but probably a better approach is to NIL the item as next described, unless there is insufficient space.

Where it is required to omit dimensions which have previously been recorded the easiest method is to write the word 'NIL' in the squaring column as shown in the following example.

	23.50 0.75 0.80		Exc. tr., width > 0.30 m, max. depth ≤ 1.00 m.
			&
	8.20 0.90 0.85	NIL	Fillg. to excavns. a.b.

Figured Dimensions

When taking off it is most desirable to use figured dimensions on the drawings in preference to scaling, since the drawings are almost invariably in the form of prints, which are not always true to scale. It is sometimes necessary to build up overall dimensions from a series of figured dimensions and this work must be set down in waste, on the right-hand side of the description column.

Numbering and Titles of Dimension Sheets

Each dimension sheet should be suitably headed with the title and taking off section of the project at the top of each sheet and with each sheet numbered consecutively at the bottom. Some prefer to number the columns on each dimension sheet rather than the pages. The practice of consecutive numbering ensures the early discovery of a missing sheet. A typical heading for a dimension sheet follows.

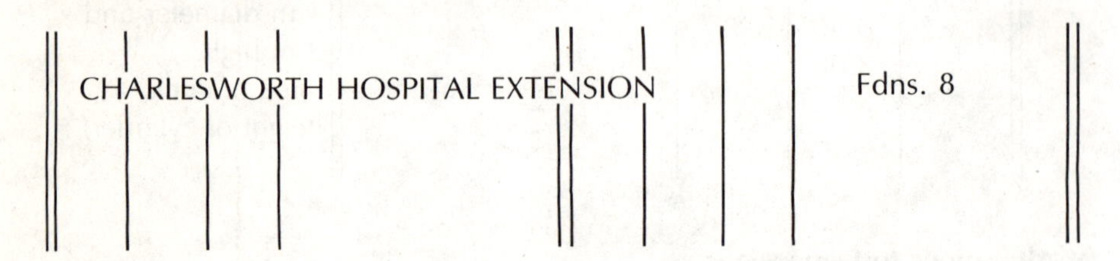

At the top of the first dimension sheet for a project it is good practice to enter a list of the drawings from which the measurements have been taken, with the precise drawing number of each contract drawing carefully recorded.

The importance of listing the contract drawings from which the dimensions have been obtained in this way, is that in the event of changes being made to the work as originally planned resulting in the issue of amended drawings, it will be clearly seen that these changes occurred after the bill of quantities was prepared and that variations to the quantities can be expected. It is in fact a *Standard Method* requirement to include in the Preliminaries Bill a list of the drawings from which the bill of quantities was prepared (SMM A11.1.1).

It is good practice to hole all dimension sheets at their top left-hand corner and to fasten together with treasury tags.

Order of Taking Off

The order of taking off largely follows the order of construction to simplify the work and to reduce the risk of items being missed, but it is not necessarily that adopted in SMM7. The measured items will subsequently be sorted into bill order which can embrace the work sections in SMM7, to secure uniformity and assist with computerisation. For instance, foundation work will be spread over a number of SMM work sections, such as D20 (Excavating and Filling), E10 (*In situ* concrete), E20 (Formwork for *in situ* concrete), E30 (Reinforcement for *in situ* concrete), E41 (Worked finishes/cutting to *in situ* concrete), F10 (Brick/block walling) and F30 (Accessories/sundry tests for brick/block/stone walling). In a simple building the order of taking off could take the form shown in the following schedule, although it will be appreciated that this may be varied to suit individual preferences and specific locations.

Sections of Work	*Broad Classifications*
1. Foundation work up to and including damp-proof course; 2. Brickwork, including facework; 3. Blockwork; 4. Fireplaces, chimney breasts and stacks; 5. Floors (solid and suspended); 6. Roofs (pitched and flat, including coverings and rainwater installation);	Carcass (structure)
7. Windows, including adjustment of openings; 8. Doors, including adjustment of openings; 9. Fittings and fixtures; 10. Stairs; 11. Finishings (walls, ceilings and floors); 12. External works, including roads, paths, fences and grassed areas; 13. Drainage work; 14. Plumbing installation; 15. Other services.	Finishings and Services

Adjustment of Openings and Voids

When measuring areas of excavation, concrete oversite, brickwork and blockwork, the most convenient practice is usually to measure the full area in the first instance, and to subsequently adjust for any voids or openings. The adjustments for the brickwork and finishings to the window and door openings are usually taken at the same time as taking off the windows or doors. This is a more logical and satisfactory method of measuring and results in a smaller overall error occurring if the very worst happens and a window or door is inadvertently omitted from the dimensions.

Descriptions

General Requirements
Considerable care and skill are required to frame adequate, and yet at the same time, concise descriptions. This is probably the most difficult aspect of taking off work and one which the student should take great pains to master. The vetting of descriptions forms an important part of editing the bill.

In addition to covering all the matters detailed in the *Standard Method of Measurement of Building Works*, the descriptions must include all the information which the estimator will require to build up a realistic price for the item in question. Where there is doubt in the mind of the estimator as to the full nature and/or extent of the item being priced, then the description is lacking in some essential feature. Descriptions can often be shortened significantly by references to clauses in the project specification as described in chapter 1, and use may be made of the SMM7 *Library of Standard Descriptions*.

Order and Form of Wording
The first few words of a description should clearly indicate the nature of the item being described. The description is badly worded if the reader has to wait almost to the end of the description to determine the subject of the item. The following example serves to illustrate this point and the first type of description is sometimes produced by students when commencing their studies in this subject.

'Bit. felt, lapped 100 mm at jts. b. & p. in ct. and laid on 102 mm bk. walls, with a width not exceeding 225 mm as dpc.'

This description would be far better worded as follows. 'Dpc, width ≤ 225, hor., single layer of hessian base bit. felt, to BS 743 type A & bedded in c.m. (1:3).'

The second description indicates at the outset the nature of the item under consideration, including the width, range and plane, in which the damp-proof course is to be laid in accordance with SMM F30.2.1.3.0 followed by a full description of the materials used as listed in SMM F30.S4–6. It will further be noted from SMM F30.C2 that pointing the exposed edges of damp-proof courses is deemed to be included and does not require specific mention, and that no allowance is made for laps (SMM F30.M2).

The use of a hyphen between two dimensions in a description, such as 150–300, shall mean a range of dimensions exceeding the first dimension stated but not exceeding the second (SMM General Rules 4.4). A dimensioned description for an item shall define and state all the dimensions necessary to identify the shape and size of the work (SMM General Rules 4.7).

Practical Implementation of Description Preparation

The wording of billed descriptions can vary considerably and it is possible to interpret and implement the provisions of SMM7 in differing ways. For instance, the author has used the terms thickness and width followed by the appropriate dimensions and excluding the mm symbol, whereas other surveyors may prefer to use expressions such as 150 mm thick and 50 mm wide following past practice. The main advantages to be gained by adopting the approach used in this book are that it conforms more closely to the wording of SMM7, permits greater rationalisation, facilitates computerisation, and is similar to the method used in the measurement and description of civil engineering work based on the *Civil Engineering Standard Method of Measurement*, thereby securing increased uniformity in the description of measurement of all types of construction work. Similarly, some surveyors may prefer to use the traditional terms 'not exceeding' and 'exceeding' instead of the symbols $\leq$ and $>$. However, these symbols are used throughout SMM7, have the merit of brevity and clarity and will, in the opinion of the author, soon gain general recognition and usage.

It will be apparent that there will, in practice, be a variety of different methods adopted for framing billed descriptions, despite the extensive work undertaken by the Building Project Information Committee and the sponsoring bodies, and the wealth of published integrated documentation described in chapter 1. It is anticipated that many architects' drawings and specifications will continue to be prepared without reference to the codes of procedure for production drawings and project specifications and the national specifications, and that many quantity surveyors will tend to follow their own personal preferences with regard to bill preparation, so that one universal procedure is unlikely to emerge. Furthermore, the preface to SMM7 permits some flexibility in writing bills of quantities and does not prohibit the use of standard prose.

The student may find all this rather bewildering but must not lose sight of the prime objective: namely, to produce bills of quantities which fully and accurately represent the quantity and quality of the works to be carried out, founded on a uniform basis for measuring building works emanating from SMM7 and embodying the essentials of good practice as defined in SMM General Rules 1.1.

Number of Units

In some cases it is necessary to give the number of units involved in a superficial or linear item, in order that the estimator can determine the average area or length of unit being priced. For instance SMM L40.1.1.1.0 requires the number of panes of glass, not exceeding 0.15 m^2, to be included in the description of the item as indicated in the following example.

3/6/	0.20 0.32	Glazg. w.stand. plain glass, in panes area ≤ 0.15 m², c.s.g. (OQ) 3 th. to BS 952 & glzg. to wd. w l.o. putty & sprgs. (In 18 nr panes)

Measurement of Similar Items

Where an entry on the dimensions paper is to be followed immediately by a similar item, the use of the words 'ditto' or 'do.', meaning 'that which has been said before', will permit the description of the next item to be reduced considerably, for example

2/4/	0.40 0.65	Ditto in panes, 0.15 – 4.00 m², do.

Another practice is to use the expressions 'a.b.' (as before) or 'a.b.d.' (as before described), to refer to a description which has occurred at some earlier point in the taking off. Care must be taken in the use of both 'ditto' and 'a.b.' to ensure that no misunderstanding as to meaning and content can occur.

Extra Over Items

When measuring certain types of work they are described as being extra over another item of work which has been previously measured. The estimator will price for the extra or additional cost involved in the second item as compared with the first. A typical example is the measurement of rainwater pipe and gutter fittings as extra over the cost of the pipe or gutter in which they occur, and which has been measured over the fittings.

Deemed to be Included Items

In the SMM the expression deemed to be included is used extensively and indicates that this particular work is covered in the billed item without the need for specific mention. It is essential that the estimator is fully aware of all these items since he must include for them when building up the unit rates.

Typical examples are all rough and fair cutting which is deemed to be included in brickwork and blockwork (SMM F10.C1b), roof coverings in slates or tiles are deemed to include underlay and battens and work in forming voids $\leq 1.00 \text{ m}^2$ other than holes (SMM H60.C1&M1), while excavating drain trenches is deemed to include earthwork support, consolidation of trench bottoms, trimming excavations, filling with and compaction of general filling materials and disposal of surplus excavated materials (SMM R12.C1).

Accuracy in Dimensions

It is essential that all dimensions shall be as accurate as possible as inaccurate dimensions are worthless. A generally accepted limit of permissible error is around 1 per cent based on full working drawings, and so the student must exercise the greatest care in arriving at dimensions. Work in waste calculations should be to the nearest millimetre.

Use of Schedules

When measuring a number of items with similar general characteristics but of varying components, it is advisable to use schedules as a means of setting down all the relevant information in tabulated form. This materially assists the taking off process and reduces the liability of error. It would be a very lengthy process indeed to take off each item in detail separately and would involve the repetition of many similar items.

The use of schedules is particularly appropriate for the measurement of a considerable number of doors, windows or manholes, a number of lengths of drain and internal finishings to a series of rooms with different finishings. An example of a schedule of internal finishings is given in chapter 9 and a manhole schedule in appendix IV. In some instances schedules are used to collect together specification information to assist in speedy taking off, while in other instances schedules are used for recording measurements and are in effect the taking off. An example of the latter would be a drainage run schedule.

Query Sheets

When taking off in practice the quantity surveyor will enter any queries for the architect on query sheets, normally divided down the centre to accommodate the queries on the left-hand side of the sheet and the answers on the right-hand side. In the examination the candidate will often have to decide the queries as they arise, when it will be desirable for him to indicate briefly in waste why he has adopted a certain course of action, and where appropriate to prepare a query sheet adopting a similar approach to that used in the office.

Preambles

Preambles are clauses usually inserted at the head of each work section bill and principally contain descriptions of materials and workmanship, as found in specifications, together with any other relevant information of which the contractor should be aware in pricing. In practice the full requirements of the SMM are frequently not given in descriptions and the remaining information is contained in preambles, thus reducing the length of billed item descriptions. Indeed much of the information frequently found in preambles is of a specification type, and there is a distinct advantage in inserting it in the bill of quantities which is always a contract document, whereas the specification is probably not. Where bill items include the term 'as described' this often means as described in preambles. Many government contracts use a separate specification document which largely replaces the preambles, and this does have some advantages for site management. Project specifications can either be separate documents or written into bills of quantities as preamble clauses. Some typical preamble clauses are given in chapter 16.

Another procedure which has been used on occasions is to combine all preamble clauses in a separate bill, following preliminaries. The contents of the preambles bill are often extracted from sets of standard clauses, such as those prepared by the former Greater London Council in *Preambles to Bills of Quantities*.

Prime Cost Items

The term 'prime cost sum' (often abbreviated to pc sum) is a sum provided for works or services to be executed by a nominated sub-contractor, or for materials or goods to be obtained from a nominated supplier. Such sums are exclusive of any profit required by the main contractor and provision is made for its addition following the pc sum.

Thus the term includes specialist work carried out by persons other than the main contractor and for materials or components to be supplied to him by persons nominated by the architect. A typical example of a prime cost item follows.

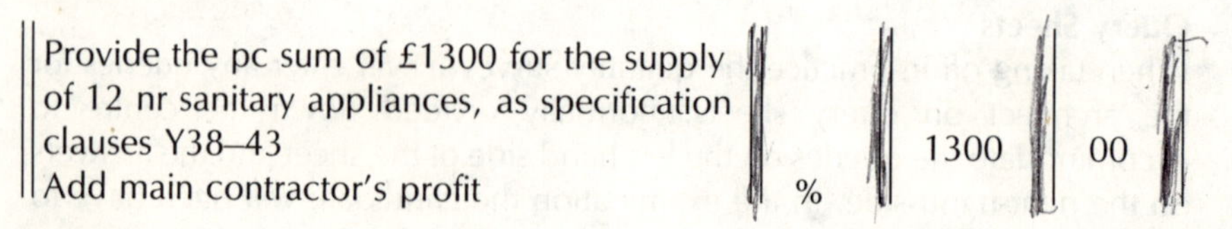

Description			£	p
Provide the pc sum of £1300 for the supply of 12 nr sanitary appliances, as specification clauses Y38–43			1300	00
Add main contractor's profit		%		

General attendance by the main contractor on nominated sub-contractors is provided as an item in the Preliminaries Bill (SMM A42.1.16.1.0). Special attendance items required by the sub-contractor, such as scaffolding, hardstandings, storage and power, are also inserted in the Preliminaries Bill. These may be entered either as fixed or time related charges, depending on whether the costs are incurred at a specific time or whether they are spread over a period (SMM A51.1.3.1–8.1–2).

Provisional Sums

Where the work cannot be described and given in accordance with SMM rules, it shall be given as a provisional sum and identified as for either defined or undefined work. In defined work items, a description and indication of the amount of work can be given, and the contractor will be deemed to have made due allowance in programming, planning and pricing the preliminaries. Where these details cannot be supplied, the work is classified as undefined and the contractor will be deemed not to have made any allowance in programming, planning and pricing preliminaries (SMM General Rules 10.2–6).

Work by local authorities and statutory undertakings are the subject of provisional sums (SMM A53.1.1–2). An example of an undefined provisional sum, which would be inserted in the Preliminaries Bill, follows.

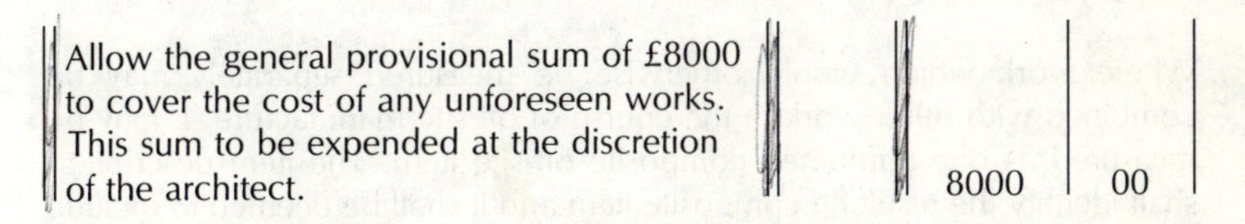

Description			£	p
Allow the general provisional sum of £8000 to cover the cost of any unforeseen works. This sum to be expended at the discretion of the architect.			8000	00

Work of Special Types

Work of each of the following special types shall be separately identified (SMM General Rules 7.1):

(a) Work on or in an existing building. Work to existing buildings is work on, in or immediately under work existing before the current project. A description of the additional preliminaries/general conditions appertaining to the work to the existing building shall be given drawing attention to any specific requirements (SMM General Rules 13).

(b) Work to be carried out and subsequently removed (other than temporary works).

(c) Work outside the curtilage of the site.

(d) Work carried out in or under water shall be so described stating whether canal, river or sea water and (where applicable) the mean spring levels of high and low water.
(e) Work carried out in compressed air shall be so described stating the pressure and the method of entry and exit.

Fixing, Base and Background

Method of fixing shall only be measured where required by the rules in each work section. Where fixing through vulnerable materials is required to be identified, vulnerable materials are deemed to include the materials defined in SMM General Rules 8.3e. Where the nature of the background is required to be identified, they shall be identified in the following classifications:

(a) Timber which shall be deemed to include manufactured building boards.
(b) Masonry which shall be deemed to include concrete, brick, block and stone.
(c) Metal.
(d) Metal faced material.
(e) Vulnerable materials which shall be deemed to include glass, marble, mosaic, tiled finishes and the like (SMM General Rules 8.1–3).

Composite Items

Where work which would otherwise be measured separately may be combined with other work in the course of off-site manufacture, it may be measured as one combined composite off-site item. The item description shall identify the resulting composite item and it shall be deemed to include breaking down for transport and installation and subsequent re-assembly (SMM General Rules 9.1).

General Definitions

Where the SMM rules require 'curved, radii stated', details shall be given of the curved work, including if concave or convex, if conical or spherical, if to more than one radius, and shall state the radius or radii. The radius is the mean radius measured to the centre line unless otherwise stated (SMM General Rules 14.1–2).

Services and Facilities

Preliminaries Bill items include such services and facilities as power, lighting, safety, health and welfare, storage of materials, rubbish disposal, cleaning, drying out, protection and security (SMM A42.1.1–15.1–2.0). There are no longer protection items in each work section as was the case with SMM6.

Plant Items

The contractor is given the opportunity in the Preliminaries Bill to price the various items of mechanical plant that generate costs which are not proportional to the quantities of permanent work. The principal items of mechanical plant are listed in SMM A43.1.1–9.1–2.0, and range from cranes and hoists to earthmoving and concrete plant.

3 Use of Mensuration in Quantities

INTRODUCTION

Mensuration is concerned with the measurement of areas and volumes of triangles, rectangles, circles, and other figures, and some basic knowledge of this subject is required by all quantity surveying students. This chapter sets out to explain how the principles of mensuration are used in the measurement of building quantities.

A list of mensuration formulae is included in appendix II for reference purposes. On the figures that follow dimensions containing a decimal marker are in metres and all others are in millimetres. Readers who are not familiar with metric dimensions may find the conversion table in appendix III helpful.

GIRTH OF BUILDINGS

Rectangular Buildings

One of the most common mensuration problems with which the quantity surveying student is concerned is the measurement of the girth or perimeter of a building. This length is required for foundations, external walls and associated items.

The length may be calculated on a straightforward rectangular building by determining the total external length of walling and making a deduction

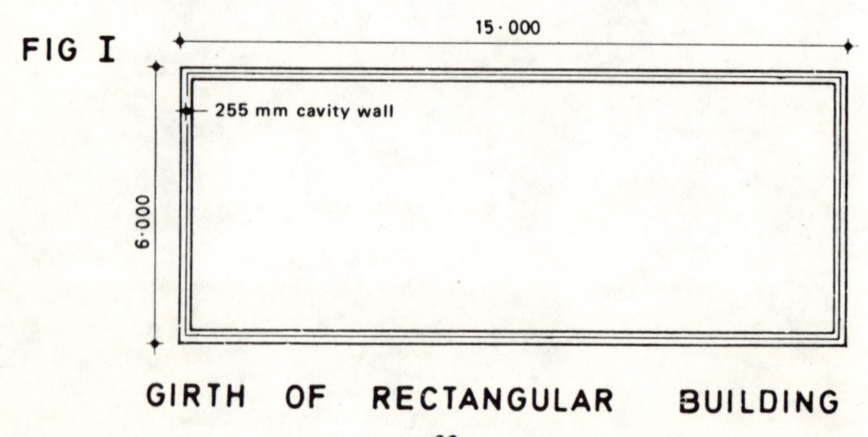

GIRTH OF RECTANGULAR BUILDING

for each external angle equivalent to the thickness of the wall. Alternatively, the internal length might be taken and an addition made for each of the external angles. The following example will serve to illustrate this point.

Taking external dimensions

	15.000
	6.000
Sum of one long and one short side	2/21.000
Sum of all four sides (measured externally)	42.000
Less corners 4/255	1.020
Mean girth of external wall (centre line of cavity)	40.980

Taking internal dimensions

	14.490
	5.490
	2/19.980
	39.960
Add corners 4/255	1.020
Mean girth of external wall	40.980

The following sketch illustrates why it is necessary to take the full thickness of the wall when adjusting for corners.

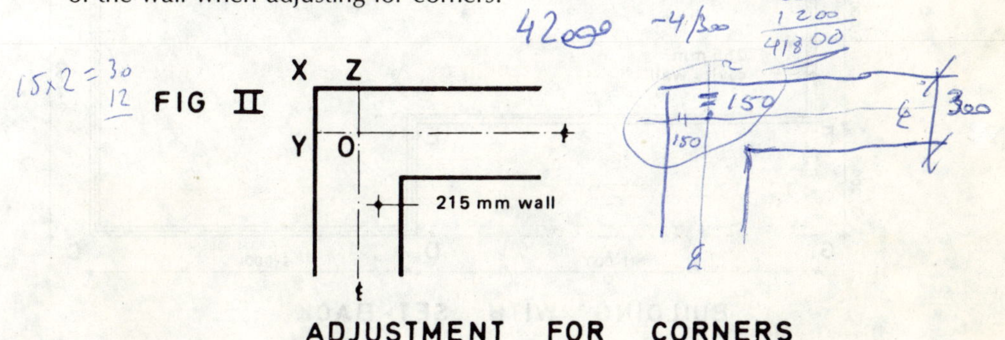

FIG II

ADJUSTMENT FOR CORNERS

Assuming that in this particular case the external dimensions have been supplied and these have to be adjusted to give the girth on the centre line through the intersection point O. The procedure is made clearer if the centre lines are extended past O to meet the outer wall faces at Y and Z. It is then apparent that the lengths to be deducted are XY and XZ, which are equal to

OZ and OY, respectively. These are equivalent to half the thickness of the wall in each case and so together are equal to the full thickness of the wall.

This centre line measurement is extremely important, since it provides the length to be used in the dimensions for trench excavation, concrete in foundations, brickwork and damp-proof course. In the case of hollow walls with a faced outer skin, three different lengths will be required when taking off the dimensions of the brickwork, since the two skins and the forming of the cavity have each to be measured separately.

The centre line of wall measurement will be required for the formation of the cavity and an addition or deduction will be needed for each corner, on a plain rectangular building, to give the centre line measurement of each skin. The amount of the adjustment for each corner will be the thickness of the skin plus the width of the cavity, that is, 102.5 mm + 50 mm = 152.55 mm on a 255 mm hollow wall.

Buildings of Irregular Outline

The position is a little more confusing if the building has an irregular outline as shown in Fig. III.

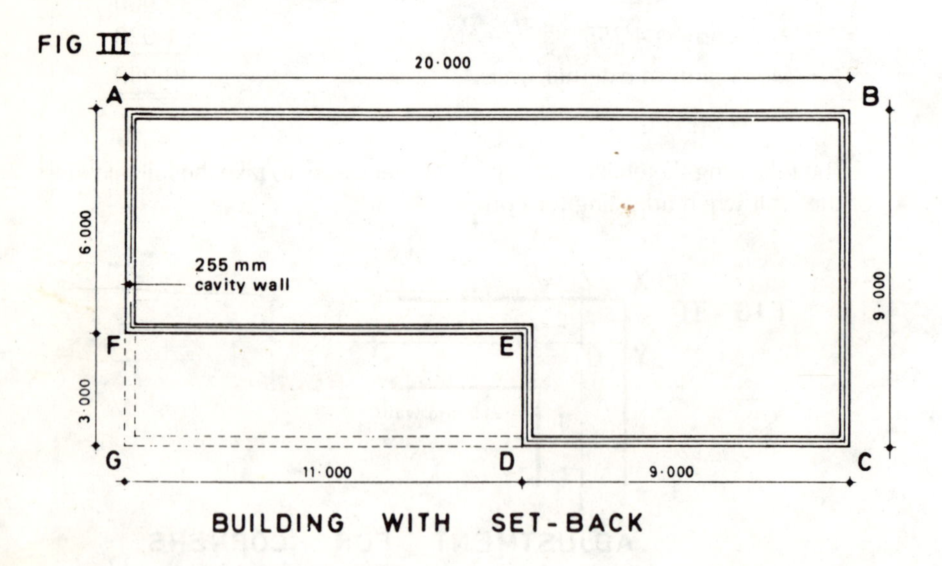

In this case the internal and external angles, E and D, at the set-back cancel each other out, and the total length is the same as if there had been no recess and the building was of plain rectangular outline (ABCG), as shown by the broken lines.

The length on centre line of the enclosing walls can be found as follows.

	20.000
	9.000
	2/29.000
	58.000
Less corners 4/255	1.020
length on centre line (℄)	56.980

On occasions buildings are planned with recesses and these involve further additions when arriving at the girth of the enclosing walls, as illustrated in the example in Fig. IV.

In this case twice the depth of the recess, that is, BC + ED, will have to be added to the lengths of the sides of the enclosing rectangle, AFGH. The internal and external angles at C, D, B and E, cancel each other out and the length of CD is equal to BE.

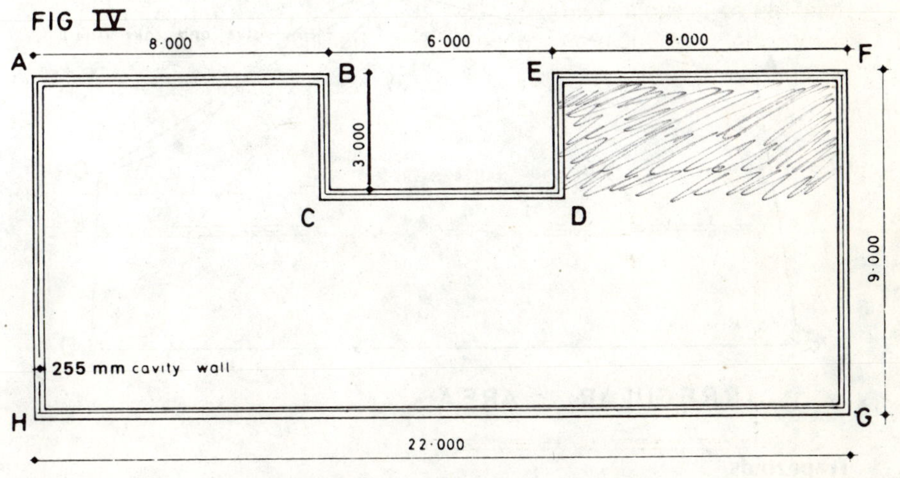

BUILDING WITH RECESS

The length of enclosing walls measured on centre line is found as follows.

	22.000
	9.000
	2/31.000
	62.000
Add twice depth of recess (2/3.000)	6.000
	68.000
Less corners 4/255	1.020
Total length of walls on ℄	66.980

MEASUREMENT OF AREAS

The quantity surveyor is often called on to calculate the areas of buildings, sites, roads, and other features, and some of the basic rules of mensuration need to be applied to the problems that arise. Some of the more common cases encountered in practice will now be illustrated.

Irregular Areas

In the measurement of irregular areas the best procedure is generally to break down the area into a number of triangles, as indicated in Fig. V by the broken lines EB and BD, giving three triangles, EAB, EBD and BCD. The area of each triangle is found by multiplying the base by half the height. Where irregular boundary lines are encountered as between B and C, the easiest method is to draw in a straight give-and-take line with a set square.

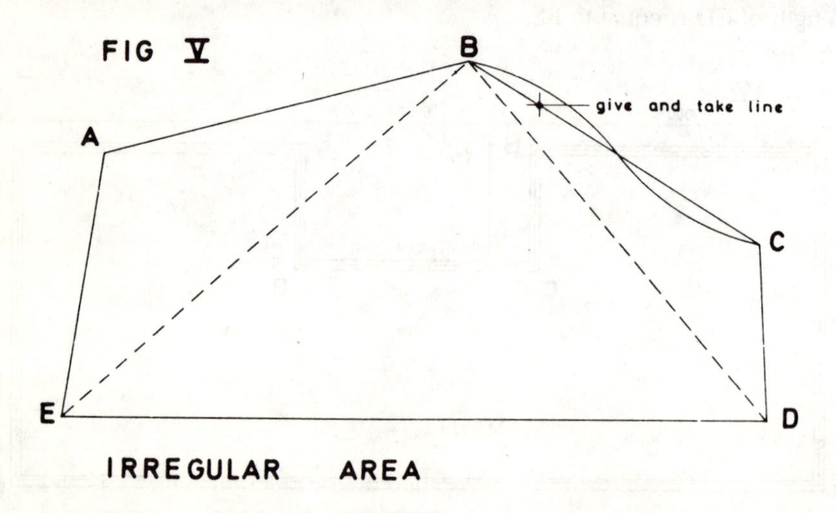

Trapezoids

Another type of irregular area which has to be measured on occasions, particularly with cuttings and embankments, is the trapezoid, which comprises a quadrilateral (four-sided figure) with unequal sides but with two opposite sides parallel. The term trapezium is often used in place of trapezoid.

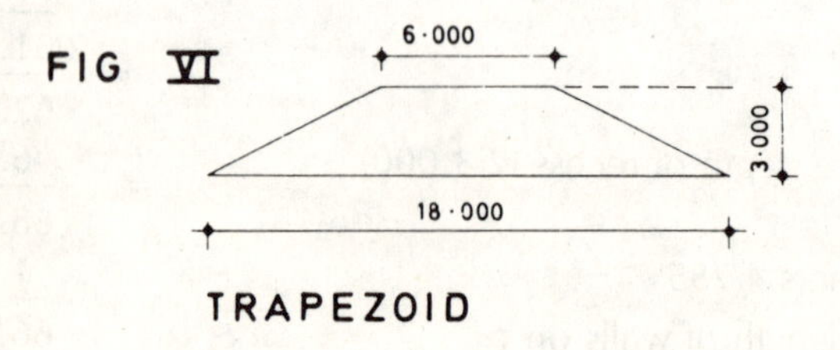

In this case the area is found by multiplying the height (vertical distance between the top and bottom of the trapezoid) by the average width, $\frac{1}{2}$(top plus bottom).

The area of the trapezoid illustrated in Fig. VI is

$$\tfrac{1}{2}(6.000 + 18.000) \times 3.000 = 12 \times 3 = 36 \text{ m}^2$$

Segments

The area of a segment (part of circle bounded by arc and chord) is sometimes required in the measurement of an arch. The normal rule for the measurement of the area of a segment is to take the area of the sector and deduct the area of the triangle, the area of the sector being found from the following formula

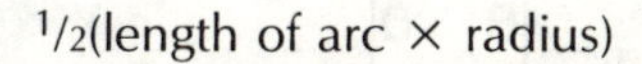

$$\tfrac{1}{2}(\text{length of arc} \times \text{radius})$$

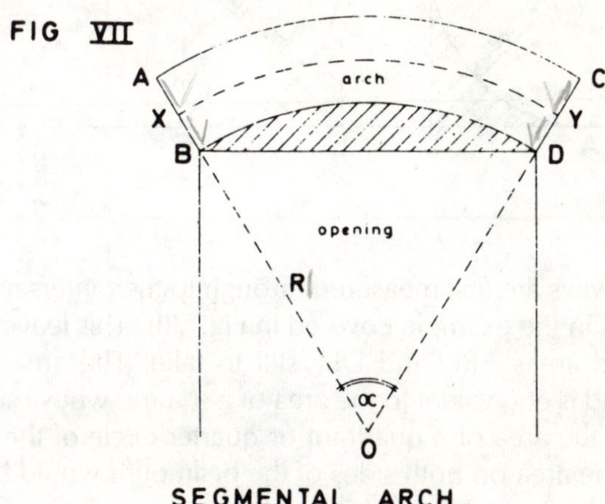

FIG VII

SEGMENTAL ARCH

The area of the segment shown hatched below the arch in Fig. VII can be determined in this way, and will be required for deduction purposes as part of the window or door opening. The length of arc BD can be found by its proportion of the circumference of the circle as a whole, related to the angle which it subtends at the centre of the circle ($<\alpha$). Thus

$$\frac{\text{arc BD}}{\text{circumference of whole circle}} = \frac{<\alpha}{360°}$$

$$\text{arc BD} = <\alpha \times 2\pi R/360°$$

Alternatively the length BD may be scaled off the drawing. The length of the arch itself will be taken on its centre line XY, since the mean length will be required. The same alternative methods of measurement are available. Another formula often used by quantity surveyors for obtaining the area of a segment follows on page 38.

$$\frac{H^3}{2C} + \frac{2}{3}CH$$

where C is length of chord and H is height of segment.

Bellmouths, as at Road Junctions
Difficulty is often experienced in measuring the irregular areas which arise at road junctions. The following example should clarify this problem.

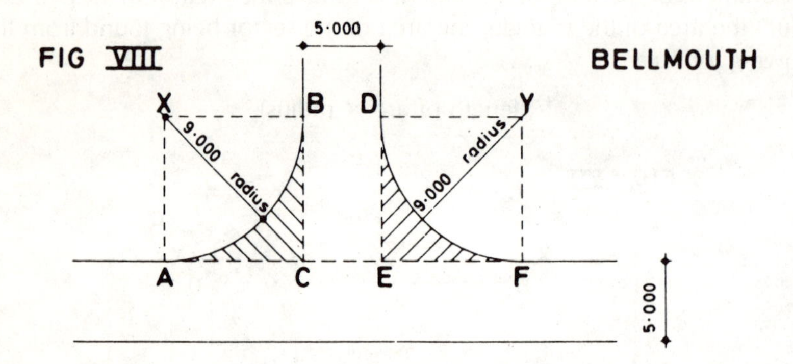

The carriageways are first measured through to their intersections for their full widths, 5 m in the example covered in Fig. VIII. This leaves the hatched irregular-shaped areas ABC and DEF still to take. The area in each case is $\frac{3}{14}$ radius^2, and is equivalent to the area of a square, whose side is equal to the radius, less the area of a quadrant or quarter circle of the same radius.

In this case the area on both sides of the bellmouth would be entered on the dimension sheet as

$2/\frac{3}{14}/$	9.00 9.00		

MEASUREMENT OF EARTHWORK

The student frequently experiences difficulty in measuring the volume of earthworks, particularly on sloping sites. The following examples are designed to indicate the main principles involved and generally clarify the method of approach.

Sloping Site Excavation

The quantity surveyor is often called on to calculate the volume of excavation and/or fill required on a sloping site and the following example indicates a comparatively simple method of approach.

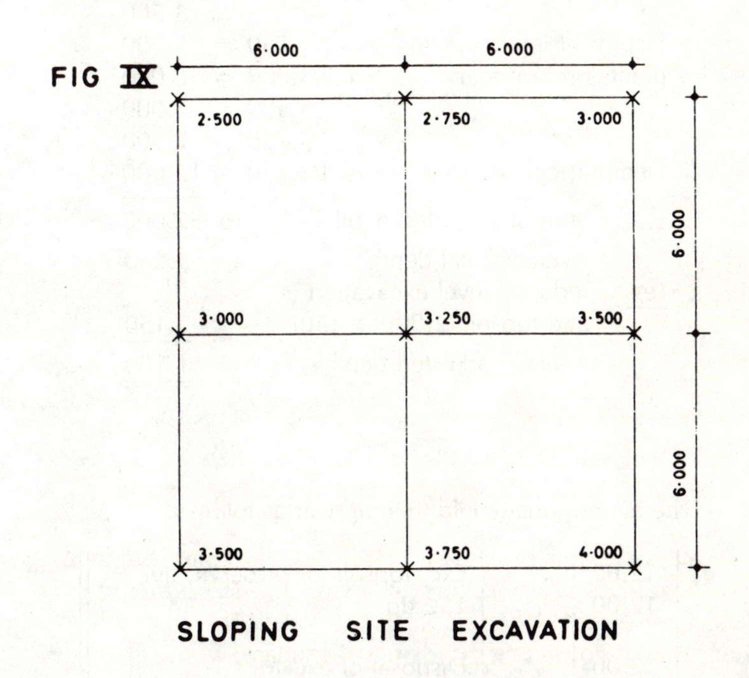

SLOPING SITE EXCAVATION

Assume that in the example illustrated in Fig. IX it is required to excavate down to a level of 2.00 m, including excavating topsoil to a depth of 150 mm. In this case the whole of the site is to be excavated, whereas if fill had been required on part of the site, it would have been necessary to have plotted the reduced level contour line on the drawing, and this line would have formed the demarcation line between the areas of excavation and fill respectively.

The average depth of excavation over the site is most conveniently found by suitably weighting the depth at each point on the grid of levels, according to the area that each level affects. This involves taking the depths at the extreme corners of the area once, intermediate points on the boundary twice and all other intermediate points four times. The sum of the weighted depths is divided by the total number of weightings (number of squares × 4) to give the average weighted depth for the whole area. This method can only be used when the levels are spaced the same distance apart in both directions.

The volume in this example is now calculated.

Corner depths		2.500
		3.000
		4.000
		3.500
Depths at intermediate points on boundary	2/2.750 =	5.500
	2/3.000 =	6.000
	2/3.500 =	7.000
	2/3.750 =	7.500
Depth at centre point	4/3.250 =	13.000
Sum of weighted depths		16)52.000
Average total depth		3.250
Less reduced level excavation and topsoil (2.000 + 150)		2.150
Average adjusted depth		1.100

The dimensions would then appear as follows.

	12.00 12.00		Exc. topsoil for preservn. av. 150 dp.
	12.00 12.00 0.15		Disposal of excvtd. mat. on site in spoil heaps av. dist. of 20.00 m from excavn.
	12.00 12.00 1.10		Exc. to red. levs., max. depth ≤ 2.00 m. & Disposal of excvtd. mat. off site.

Cuttings and Embankments

The volumes of cuttings and embankments are generally calculated from the cross-sectional areas taken from plotted cross-sections, often prepared at 30 m intervals along the line of the cutting or embankment. Certain intermediate cross-sectional areas are often weighted by using Simpson's rule or the prismoidal formula.

Furthermore, allowance must be made for the sloping banks on either side as illustrated in the following example. The bank slopes may be described as say 1 in 2 or 2 to 1, indicating that the bank rises 1 m vertically for every 2 m in the horizontal plane.

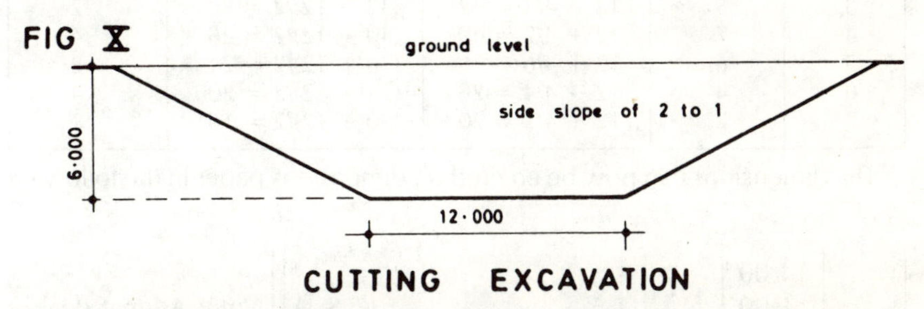

CUTTING EXCAVATION

NOTE: THIS FIGURE IS DRAWN TO A NATURAL SCALE, WHEREAS IN PRACTICE CROSS-SECTIONS ARE OFTEN DRAWN TO AN EXAGGERATED VERTICAL SCALE

When calculating the volume of excavation and fill for cuttings and embankments, Simpson's rule can often be used to advantage and a simple example follows to illustrate this point.

Using Simpson's rule the area at intermediate even cross-sections (nrs. 2, 4, 6, etc.) are each multiplied by 4, the areas at intermediate uneven cross-sections (nrs. 3, 5, 7, etc.) are each multiplied by 2 and the end cross-sections taken once only. The sum of these areas is multiplied by 1/3 of the distance between the cross-sections to give the total volume. To use this formula it is essential that the cross-sections are taken at the same fixed distance apart and that there is an odd number of cross-sections (even number of spaces, between cross-sections).

For instance, take a cutting to be excavated for a road, 180 m in length and 12 m in width, to an even gradient, with mean depths calculated at 30 m intervals as indicated and side slopes of 2 to 1.

Cross-section	1	2	3	4	5	6	7
Mean depth (m)	1	3	5	7	6	4	2

The width at the top of the cutting can be found by taking the width at the base, that is, 12 m and adding 2/2/the depth to give the horizontal spread of the banks (the width of each bank being twice the depth with a side slope of 2 to 1).

Cross-section	*Depth* (m)	*Width at Top of Cutting* (m)	*Mean Width* (m)	*Weighting*
1	1	12 + 4/1 = 16	(16 + 12)/2 = 14	1
2	3	12 + 4/3 = 24	(24 + 12)/2 = 18	4
3	5	12 + 4/5 = 32	(32 + 12)/2 = 22	2
4	7	12 + 4/7 = 40	(40 + 12)/2 = 26	4
5	6	12 + 4/6 = 36	(36 + 12)/2 = 24	2
6	4	12 + 4/4 = 28	(28 + 12)/2 = 20	4
7	2	12 + 4/2 = 20	(20 + 12)/2 = 16	1

The dimensions can now be entered on dimensions paper in the following way.

Timesing	Dimensions	Squaring	Description	Side note
	14.00 1.00		(C.S.1)	*Note:* A great deal of laborious and unnecessary labour in squaring has been avoided by entering all the dimensions as superficial items, to be subsequently cubed by multiplying the sum of the areas by 1/3 of the length between the cross-sections. (Total weighting is 18 and number of 30 m long sections of excavation is 6, so that 6/18 or 1/3 of the distance of 30 m must be the timesing factor required.)
4/	18.00 3.00		Exc. to red. levs. max. depth ≤ 8.00 m. & (C.S.2)	
2/	22.00 5.00		Disposal of excvtd. mat on site av. dist. of 50 m from excavn. & (C.S.3)	
4/	26.00 7.00		Fillg. to make up levs. av. thickness > 0.25 m arisg. from excavns. (C.S.4)	
2/	24.00 6.00		Cube × $\frac{1}{3}$/30.00 (C.S.5)	
4/	20.00 4.00		(C.S.6)	
	16.00 2.00		(C.S.7)	

In simpler cases involving three cross-sections only, the prismoidal formula may be used, whereby

$$\text{volume} = \frac{1}{6}\left\{\begin{matrix}\text{total}\\ \text{length}\end{matrix}\right\} \times \left\{\begin{matrix}\text{area of}\\ \text{first section}\end{matrix} + \begin{matrix}\text{4 times area of}\\ \text{middle section}\end{matrix} + \begin{matrix}\text{area of last}\\ \text{section}\end{matrix}\right\}$$

These formulae can also be used to calculate the volume of banks which vary in cross-sectional area throughout their lengths.

MEASUREMENT OF PITCHED ROOFS

Lengths of Rafters

Where roof sections are drawn to a sufficiently large scale the easiest method is to scale the length of the rafter off the drawing, taking the length from one extremity to the other of the rafter.

Another alternative is to calculate the length by multiplying the natural secant of the angle of pitch by half the total span of the roof. The natural secants of the more usual pitches of roof are as follows.

Pitch of roof	15°	30°	40°	45°	50°
Natural secant	1.036	1.155	1.305	1.414	1.555

The student is referred to four-figure mathematical tables for values of natural secants relating to other angles of pitch. The following example illustrates the method of calculation of the lengths of rafters.

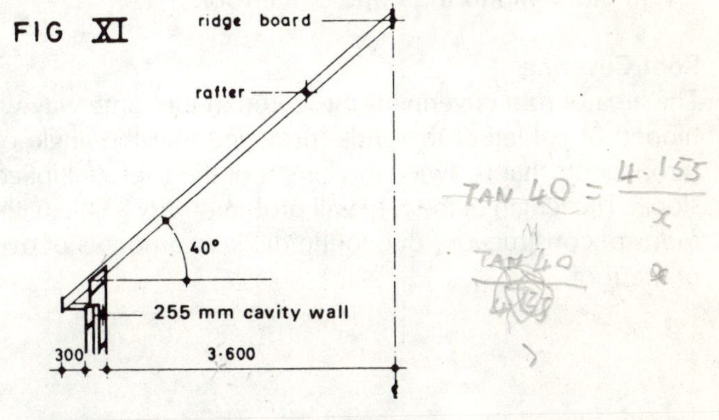

LENGTHS OF RAFTERS

Half total span of roof =	3.600 (half effective span)	+ 255 (wall thickness)	+ 300 (overhang at eaves)	= 4.155

Length of rafter = 4.155 × 1.305 (secant 40°)
= <u>5.422 m</u>

(to which a small addition of 75 mm should be made for a tapered end to be precise).

Lengths of Hips and Valleys

The length of a hip or valley is most conveniently found by plotting and scaling from the roof plan as shown in Fig. XII.

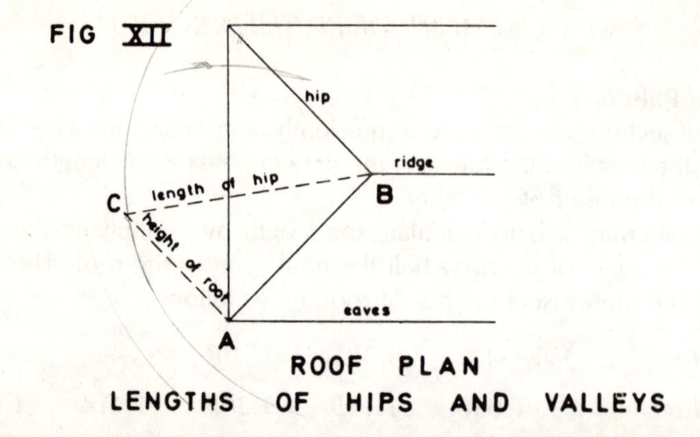

ROOF PLAN
LENGTHS OF HIPS AND VALLEYS

The length AB represents the length of the hip on plan, while the length on slope is actually required. To obtain the length on slope, the height of the roof is set out at right angles to AB on the line AC. BC then represents the length of the hip to the slope of the roof.

Roof Coverings

The area of roof covering is measured in the same way whether the roof be hipped or gabled at the ends, provided that the angle of pitch is constant throughout, that is, twice the length of the roof multiplied by the length on slope. The length of the roof will probably vary a little with each of these two forms of construction, due to the different amounts of overhang at the ends of the roof.

4 Measurement of Excavation and Foundations

Preliminary Investigations

Before taking off any dimensions the quantity surveyor normally makes a careful study of all the drawings relating to the project to obtain the overall picture and become familiar with the main details. The next step is usually a visit to the site to obtain details and measurements of any work required on the site, often termed siteworks. These works include breaking up paving, taking down boundary walls and fences, felling trees, grubbing up hedges, and similar work.

Some contracts involve alterations to existing buildings and these are sometimes termed 'spot items' and are normally kept together in a separate section of the bill of quantities possibly headed 'Demolitions and Alterations', with some of the items covered by Provisional Sums, where the full extent of the work involved cannot be accurately determined. Many of the details relating to spot items will also need to be obtained on the site.

When visiting the site the quantity surveyor should also be on the look-out for any unusual items which affect cost, and which should accordingly be included in the billed description. A check on the type of soil and groundwater level comes into this category, unless the information is supplied by the architect, probably in the form of particulars from trial pits or boreholes excavated on the site.

General Items

It is necessary to give details and location of any trial pits or boreholes, the groundwater level on the site (at a prescribed pre-contract date), and details and location of existing services. Alternatively, in the absence of any trial pits or boreholes, a description of the ground and strata to be assumed should be given. Information should also be provided on any features on the site which are to be retained (SMM D20.P1).

Site Preparation

Removal of trees and tree stumps is measured as an enumerated item, including grubbing up roots, disposal of materials and filling voids, classified in the girth ranges listed in SMM D20.1.1–2.1–3. Tree girths are measured at a height of 1.00 m above ground and stump girths at the top (SMM D20.M1–2). Clearing site vegetation is measured in m^2 with a description sufficient for identification purposes (SMM D20.1.3.4.0). Site vegetation embraces bushes, scrub, undergrowth, hedges, trees and tree stumps ≤ 600 mm girth (SMM D20.D1).

The first building operation is normally the excavation of the topsoil for preservation over the whole area of the building and this usually forms the first excavation item in the Excavating and Filling section of the bill of quantities. The area is measured to the outer extremities of the foundations in m^2 and the average depth, often 150 mm, is included in the description (SMM D20.2.1.1.0). Disposal of topsoil on the site in temporary spoil heaps for re-use is covered by a separate cubic item stating the location of the spoil heaps (SMM D20.8.3.2.1). Spreading soil on the site to make up levels is measured in m^3, distinguishing between thicknesses ≤ and > 0.25 m. Disposal of excavated material off the site is measured in m^3, giving details of specified locations or handling where appropriate (SMM D20.8.3.1.1–2). If the existing turf over the site of the building is to be preserved, then this forms a separate billed item measured in m^2, stating the method of preserving the turf (SMM D20.1.4.1.0).

Excavation to Reduce Levels

Where the site is sloping then further excavation is required to reduce the level of the ground to the specified formation level; this excavation is measured in m^3 as excavation to reduce levels in accordance with SMM D20.2.2.1–4.0, giving the appropriate maximum depth range. The excavation rules in SMM7 are based on all excavation being carried out by mechanical plant.

Excavation of Foundation Trenches

Foundation trench excavation is measured in m^3, stating the commencing level where > 0.25 m below existing ground level and the maximum depth range in accordance with SMM D20.2.5.1–4.1, namely ≤ 0.25 m, 1.00 m, 2.00 m and thereafter in 2.00 m stages. It is necessary to distinguish between trenches ≤ 0.30 m wide and those > 0.30 m wide.

All excavation is measured net with no allowance for increasing in bulk after excavation or for the extra space required for working space or to accommodate earthwork support (SMM D20.M3). Breaking out rock; concrete; reinforced concrete; brickwork, blockwork or stonework; shall each be described and measured separately in m^3 as extra over any types of excavating (SMM D20.4.0.1–4.1), while breaking out existing hard pavings is measured in m^2, stating the thickness, as extra over excavating (SMM D20.5.0.5.1). Rock is defined as any material which is of such size or position that it can only be removed by wedges, special plant or explosives

(SMM D20.D3). Excavating below groundwater level is given in m^3 as extra over any types of excavating (SMM D20.3.1.0.0).

Working allowance to excavations, categorised in four types of excavation as SMM D20.6.1–4, is measured in m^2, where the face of the excavation is $<$ 600 mm from the face of formwork, rendering, tanking or protective walls (SMM D20.M7).

Excavating next to existing services is measured in metres as extra over any types of excavating, stating the type of service (SMM D20.3.2.1.0). While that around existing services crossing excavation is an enumerated extra over item (SMM D20.3.3.1.0), since it is likely to entail hand digging, whereas most other excavation will be carried out by machine.

Three sets of levels will be required before foundation work can be measured: (1) bottom of foundations, (2) ground levels and (3) finished floor levels.

When measuring foundation trenches it is advisable to separate the trenches to external and internal walls. Where the external wall foundation is of constant width and the site is level, its measurement presents no real difficulty with the length being obtained by the normal girthing method, as outlined in chapter 3. Where the site is sloping and stepped foundations are introduced the process of measurement of the foundation trench excavation is more complex, since each length of trench will have to be dealt with separately. It is good policy to use a schedule for this purpose giving the lengths, levels, average depths within maximum depth ranges, and widths of trench for each section between steps. The sum of the lengths of all the individual sections will need to be checked against the total calculated length to avoid any possibility of error. Furthermore, the individual brickwork lengths will not always coincide with the lengths of excavation and concrete, as illustrated in *Advanced Building Measurement*.

After measuring the excavation for external wall foundations, the internal wall foundations will be taken, and this will often involve a number of varying lengths and widths of foundations, which are best collected together in waste and the drawing suitably marked as each length is extracted. Care must be taken to adjust for the overlap of trenches at the intersection of the external and internal walls, as shown in Fig. XIII. The external wall foundation trench will have been measured around the whole building in the first instance, and the hatched section will have to be deducted at each intersection when arriving at the length of internal wall foundation trench.

Disposal of Excavated Material

The subsequent disposal of excavated material forms a separate billed item in m^3, either of soil to be stored on site, used as filling to make up levels, filling to excavations, or to be removed off the site. In the first instance, when measuring the trench excavation, it is usual to take the full volume as filling

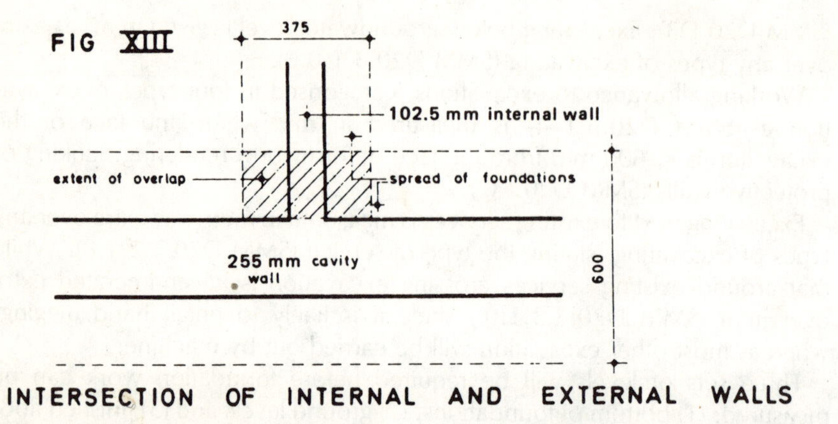

to excavations and subsequently to adjust as disposal of excavated materials off site with the measurement of the concrete and brickwork (see Fig. XIV).

This normally represents the simplest and most convenient method of approach but circumstances do arise when it is more convenient to take off the full volume as remove from site in the first instance, such as when the top of the foundation is in line with the stripped level.

Handling of excavated material is normally at the discretion of the contractor.

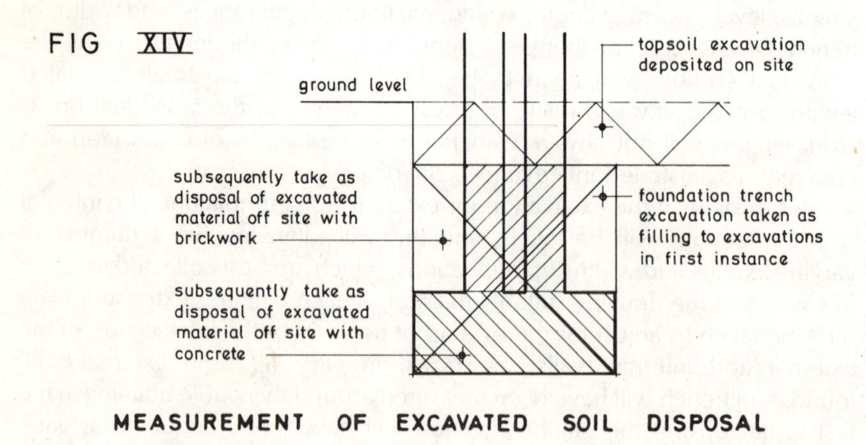

Surface Treatments

The measurement of the excavation is generally followed by a superficial item for compacting the bottom of the excavation (SMM D20.13.2.3.0). Surface treatments may alternatively be given in the description of any superficial item (SMM D20.M17). Compacting is deemed to include levelling and grading to falls and slopes$\leq$ 15° from horizontal (SMM D20.C5).

Basement Excavation

Basement excavation is measured to the outside of foundations in m^3 stating the maximum depth range. Working space allowance is measured in m^2 on the external face of the basement walls, where the face of the excavation is $<$ 600 mm from the face of the tanking or protective walls (SMM D20.M7). The area measured is calculated by multiplying the girth of the tanking or protective walls by the depth of excavation below the commencing level of the excavation (SMM D20.M8). Additional earthwork support, disposal, backfilling, work below groundwater level and breaking out are deemed to be included (SMM D20.C2).

It is usual to take the basement excavation as disposal of excavated material off site in the first instance and to later adjust the filling to excavation where appropriate.

Earthwork Support

Support to the sides of excavation is measured in m^2 to trenches (excluding pipe trenches where it is deemed to be included in the linear trench excavation description), pits, and the like, where $>$ 0.25 m in depth, whether the support will actually be required on the project or not. The maximum depth is given in stages in accordance with SMM D20.7.1–3. Earthwork support is also classified by the distance between opposing faces in stages$\leq$ 2.00 m, 2.00 to 4.00 m, and $>$ 4.00 m. Earthwork support left in, curved, next to roadways or existing buildings, below groundwater level or to unstable ground shall be described and separately measured in accordance with the rules contained in SMM D20.6–8.

Concrete Foundations

Concrete particulars are to include the kind and quality of materials, mix details, tests of materials and finished work, methods of compaction and curing and other requirements (SMM E10.S1–5), but much of this information may be included in preamble clauses or cross-references to project specification clauses. Concrete poured on or against earth or unblinded hardcore shall be so described (SMM E10.1–8.1–3.0.5). Concrete foundations include attached column bases and attached pile caps, while isolated foundations include isolated column bases, isolated pile caps and machine bases. Beds include blinding beds, plinths and thickening of beds, while slabs include attached beams and beam casings whose depth is $\leq$ three times their width (depth measured below slab) and column drop heads (SMM E10.D1–4).

In situ concrete is measured in m^3 and the degree of difficulty in placing the concrete is reflected by giving the thickness ranges $\leq$ 150 mm, 150–450 mm and $>$ 450 mm in the case of beds, slabs and walls.

On a sloping site the concrete foundations will probably be stepped and it will be necessary to measure the additional concrete at the step and a linear item of formwork to the face of the step (SMM E20.1.1.2–4) classified in three stages of depth: ≤ 250 mm, 250–500 mm and 500 mm–1.00 m.

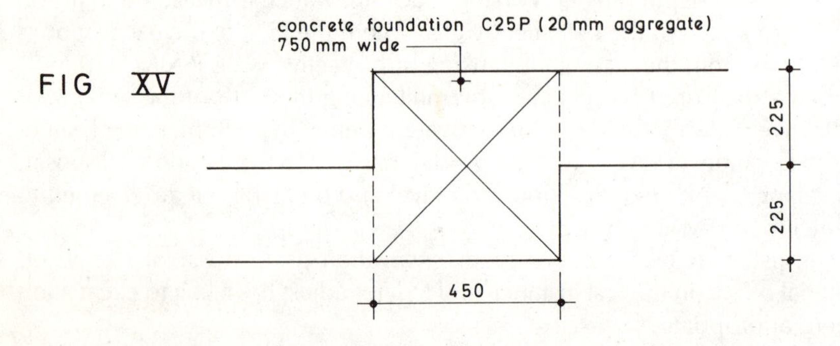

STEPPED FOUNDATION

In the example shown in Fig. XV it will be necessary to add a 450 mm length of foundation, 225 mm thick, and to make the necessary soil adjustments as shown in the following entry on the dimensions paper.

	0.75		Add *In situ* conc.	
	0.45		fdns. (C25P/20 agg.),	The symbol 'mm'
	0.23		poured on or	has been omitted
			against earth. (step	from the descrip-
			&	tions.
			Ddt. Filling to excavns.	Note the adjustment
			a.b.	of excavated soil
			&	disposal that is also
			Add Disposal of excvtd.	required.
			mat. off site.	It will also be
				necessary to take a
				750 mm length of
				formwork to sides of
				foundation, plain
				vertical, height
				≤ 250 mm.

Note: Concrete has been classified by strength, relating to a grade in CP110, as an alternative to stating the required strength, e.g. 21 N/mm^2 (20 mm aggregate) or the mix of concrete, such as 1:3:6/20 mm aggregate.

If the concrete foundations are reinforced with fabric reinforcement, the reinforcement is measured net in m^2 stating the mesh reference, weight per m^2, minimum laps and strip width, where placed in one width, (SMM E30.4.1.0.2). Bar reinforcement is billed in tonnes, keeping each diameter (nominal size) separate, although it will be entered by length on the dimension sheet, distinguishing between straight, bent and curved bars (SMM E30.1.1.1–3). Hooks and tying wire, and spacers and chairs which are at the discretion of the contractor are deemed to be included (SMM E30.C1).

Other Substructure Work

It has been customary to include all substructure work in a separate section of the bill, including brickwork up to and including damp-proof course, which may be subject to re-measurement. However, SMM7 in line with *Common Arrangement of Work Sections for Building Works*, subdivides this work into several work sections: D20 (Excavating and Filling), E10 (*In situ* concrete), E20 (Formwork for *in situ* concrete), E30 (Reinforcement for *in situ concrete*), F10 (Brick/block walling) and F30 (Accessories/sundry items for brick/block/stone walling). Hence the substructure work will probably be subdivided into these work sections in the bill of quantities, albeit rather fragmented. The measurement of brickwork is covered in detail in chapter 5, but it is considered desirable to include some brickwork in the worked examples in the present chapter to follow normal taking off practice.

Brick Walling

Brick walling is measured in m^2, stating the nominal thickness, such as 215 or one brick thick, and whether there is facework (fair finish) on one or both sides (SMM F10.1.1–3.1.0). The skins of hollow walls and the formation of the cavity, including wall ties, are each separately measured in m^2 (SMM F30.1.1.1.0).

The projecting brickwork in footings, which are no longer used very extensively, is separately measured as horizontal projections in metres, stating the width and depth of projection (SMM F10.5.1.3.0). The following example illustrates the measurement of footings.

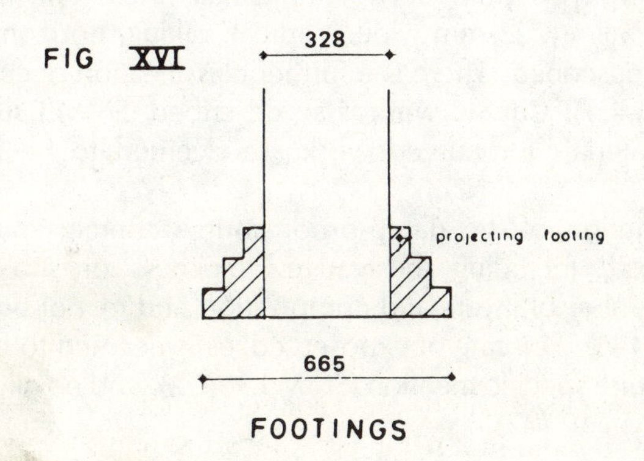

The $1\frac{1}{2}$ B wall is measured down to the base of the wall and the projections are measured as an additional linear item, taking the combined average projection on each face of the wall.

			ftgs. width of projs. on ea. face. top. cos. $\frac{1}{4}$ B bottom cos. $\frac{3}{4}$ B 2)1B av.$\frac{1}{2}$ B	
2/	10.00		Bk. projs., av. width 102 & depth 225, hor., comms. in c.m. (1:3). (assuming length of 10 m of wall)	Average combined projection measured over three courses on each face of wall.

Facework

Brick facework is included in the measurement of the brickwork on which it occurs, with a description of the kind, quality and size of bricks, type of bond, composition and mix of mortar and type of pointing (SMM F10.S1–4). In practice these particulars could alternatively be included in preamble clauses or be cross-referenced to a project specification.

Damp-proof Courses

Damp-proof courses are measured in m^2, distinguishing between those $\leq$ and $>$ 225 mm in width, for example half brick, one brick and block partitions are all $\leq$ 225 mm wide. Vertical, raking, horizontal and stepped work are so described. There is a further classification of cavity trays (SMM F30.2.1–2.1–4.1). Curved work is so described (SMM F30.M1), although the extra materials for curved work are deemed to be included (SMM F30.C1c).

The description of the damp-proof course contains particulars of the materials used, including the gauge, thickness or substance of sheet materials, number of layers and composition and mix of bedding materials (SMM F30.S4–6). Pointing of exposed edges is deemed to be included and does not require specific mention (SMM F30.C2), and no allowance is made for laps (SMM F30.M2).

WORKED EXAMPLES

Worked examples follow covering the foundations to a small building and a basement, to illustrate the method of approach in taking off this class of work and the application of the principles laid down in the *Standard Method of Measurement of Building Works* (SMM7).

The importance of a logical sequence in taking off cannot be overemphasised. It simplifies the taking off process, reduces the risk of omission of items and gains the student additional marks in the examination.

In taking off foundations to a small building a satisfactory order of items would be as follows.

(1) Excavate topsoil.
(2) Excavate foundation trench and backfill.
(3) Compact bottom of trenches.
(4) Earthwork support to trenches.
(5) Concrete in foundations, including adjustment of soil disposal.
(6) Brickwork, including cubic adjustment of soil disposal.
(7) Damp-proof course.
(8) Adjustment of topsoil outside building.
(9) Disposal of surface water.

On the drawings decimal markers in dimensions indicate measurements in metres. Where there is no decimal marker the dimensions are normally in millimetres.

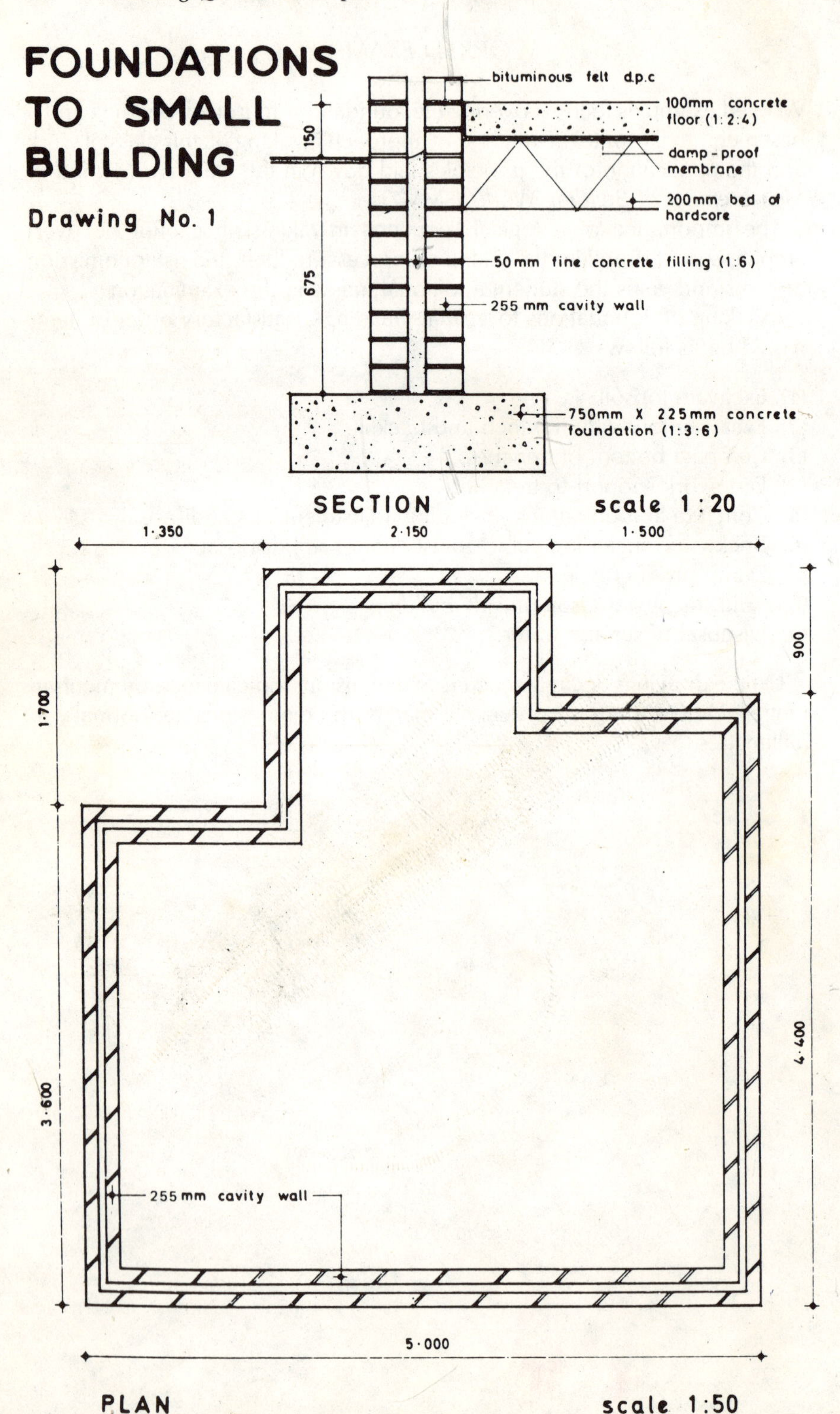
FOUNDATIONS TO SMALL BUILDING
Drawing No. 1
bituminous felt d.p.c
100mm concrete floor (1:2:4)
damp-proof membrane
200mm bed of hardcore
50mm fine concrete filling (1:6)
255 mm cavity wall
150
675
750mm X 225mm concrete foundation (1:3:6)
SECTION
scale 1:20
1·350
2·150
1·500
900
1·700
4·400
3·600
255 mm cavity wall
5·000
PLAN
scale 1:50

FOUNDATIONS TO SMALL BUILDING

Dimensions	Squared	Description	Waste
			Work up to dpc width of fdn. 750 less wall 255 2) 495 spread of fdn. 247·5
			4·400 / 900 / 5·300 / 495 / 5·795 add 5·000 fdn.sprd. 2/247·5 495 5·495
5·80 5·50	31·90	Exc. topsoil for preservn. av. 150 dp.	
1·70 1·35	2·30	Ddt. ditto.	
1·50 0·90	1·35 3·65		
5·80 5·50 0·15	4·79	Disposal of excvtd. mat. on site in spoil heaps av. dist. of 30 m from excavn.	
1·70 1·35 0·15	0·34	Ddt. ditto.	
1·50 0·90 0·15	0·20 0·54		depth 675 add fdns. 225 900 less topsoil 150 750
			len. 5·300 5·000 2/ 10·300 20·600 less corners 4/255 1·020 19·580
19·58 0·75 0·75	11·01	Exc. tr. width > 0·30 m, max. depth ≤ 1·00 m. & Fillg. to excavns., av. thickness > 0·25 m, arising from excavns.	

1.1

EXAMPLE I

Note: the dimensions in this example have been squared ready for transfer to the abstract in Example XXVI.

Note extensive use of 'waste' calculations to build-up dimensions with full descriptive notes. Metres and millimetres are used throughout in waste. The symbol 'mm' has been omitted from descriptions as it seems superfluous and no confusion should arise in practice.

Excavating topsoil to be preserved is measured in m², stating the average depth (SMM D20.2.1.1.0).

It is best to measure the rectangular area overall and then to deduct the voids from it.

Disposal is measured in accordance with SMM D20.8.3.2.1 assuming that the topsoil is to be redistributed subsequently under the contract, probably in external works, requiring further items in accordance with SMM D20.10.1.1.3.

The dimensions in the dimension column, unlike those in waste, are expressed in metres to two places of decimals (to nearest 10mm).

Note the method of building-up the depth of trench excavation and its girth on centre line.

The internal and external angles at the set-backs to the building cancel themselves out and no adjustments are necessary.

Note order of length, width and height, following the practice prescribed for dimensions in descriptions in SMM General Rules 4.1.

Excavation to trenches is measured in the prescribed stages of maximum depth, and stating the width classification as SMM D20.2.5–6.1–4.0, and the commencing level where > 0·25 m below existing ground level.

All soil excavated is taken as filling returned to trenches in the first instance (See SMM D20.9.2.1.0). The origin of the filling material shall be one of the three categories in SMM D20.9.2.1–3 and it is necessary to distinguish between filling with an average thickness ≤ or > 0·25 m.

FOUNDATIONS TO SMALL BUILDING (Contd.)

	19.58 0.75	14.69	Compactg. bott. of excavn.
			add 4/2/375 — 19.580, 3.000; outer face of tr. 22.580 less corners — 19.580, 3.000; inner face of tr. 16.580
	22.58 0.90	20.32	Earthwk. suppt., max. depth ≤ 1.00 m, distance between opposg. faces ≤ 2.00 m. (outer face
	16.58 0.75	12.44 32.76	(inner face
	19.58 0.75 0.23	3.38	In situ conc. fdns. (1:3:6/40agg), poured on or against earth. & Ddt. Fillg. to excavns. a.b.d. & Add Disposal of excavtd. mat. off site.
			bwk. ht. — 675 add above g.l. 150 825
2/	19.58 0.83	32.50	Bk. wall, thickness: 102.5, in comms. in stret. bond in c.m. (1:3).

Compacting is measured in m^2 in three separate categories (SMM D20.13.2.1-3).

To arrive at the girth of the outer face of the trench add twice times half the trench width at each of the four main corners (the other two external angles are offset by the internal ones).

Earthwork support is measured in m^2 stating the maximum depth range and the distance between opposing faces (SMM D20.7.1.1.0).
The earthwork support to the outer face of the trench will include the depth occupied by topsoil. This can be strutted from opposing face of the trench and so the distance of ≤ 2.00 m can apply.
Earthwork support is not measurable to faces ≤ 0.25 m high (SMM D20. M9a).

Note the classifications of in situ concrete in foundations as SMM E10.1. and the further requirements in SMM E10.1.0.0.5. The concrete is followed by adjustment of soil disposal. The rough finish to foundation concrete to receive walling is outside the scope of surface treatments in SMM E41.1–7.
Alternatively, the wall could be described as 'Bk wall, ½ B th'.

The part of the outer skin built in facework will be adjusted later.
Half brick skins of hollow walls are measured as half brick walls in m^2 (SMM F10.1.1.1.0 and F10. D4).
Brick dimensions would be given in the preamble to the Brickwork and Blockwork or Brick and Block Walling Bill and are normally 215 x 102.5 x 65. The description of brickwork includes bricks, bond, mortar and pointing, where appropriate (SMM F10.S1–4).

1.2

FOUNDATIONS TO SMALL BUILDING (Contd)

	19·58 0·83	16·25	Form cav. in holl. wall. width: 50, inc. 4nr. wall ties/m² of zinc. coated m.s. vert. twist type to BS 1243.	The formation of the cavity stating the width and the type, size and spacing of wall ties are given in a single superficial item (SMM F30.1.1.1.0). Note the description of the wall ties and particularly the reference to a British Standard. Alternatively, this could be cross referenced to the wall tie description in the specification or a preamble clause, and reference can be made to the SMM 7 library of standard descriptions.
	19·58 0·05 0·68	0·67	In situ conc. (1:6) fillg. holl. wall, thickness ≤ 150. 675 less topsoil 150 525	Concrete filling to cavities of hollow walls is measured in m³ with the thickness range given as SMM E10.8.1-3.
	19·58 0·26 0·53	2·70	Ddt. Fillg. to excavns. a.b.d. & Add Disposal of excavtd. mat. off site.	Adjustment of excavated soil disposal for volume displaced by brickwork below topsoil.
2/	19·58 0·10	3·92	Dpc, width ≤ 225, hor., single layer of hessian base bit. felt to BS 743 ref. A & bedded in c.m. (1:3).	Damp-proof courses are measured in m², distinguishing between those ≤ and > 225 wide, and giving the particulars listed in SMM F30.2.1-2.1-4.1, where appropriate. Pointing exposed edges is deemed to be included and does not require specific mention, and no allowance is made for laps (SMM F30.C2 and M2).
			Facewk. ht. above g.l. 150 below g.l. 75 225 len. ½ thickness of cav. (½×50) 25 ½ thickness of h.b. skin (½×102.5) 51·25 76·25 19·580 add corners 4/2/76·25 610 20·190 1.3	It is usual to allow one course of facings below ground level to counteract any irregularities in the finished ground level. Note the method of building up the girth of the outer skin, measured on its centre line, working from the mean girth of the hollow wall which has been determined previously.

FOUNDATIONS TO SMALL BUILDING (Contd.)

Timesing	Dimensions	Squaring	Description	Side notes
	20·19 0·23	4·64	Ddt Bk. wall, thickness 102·5, in comms. in c.m. (1:3). & Add Ditto facewk. o.s. in 'X' multi-col. fcg. bks. basic price £215/1000 (delvd. to site) in stret. bond in c.m. (1:3) & ptg. w.a. neat flush jt. as wk. proceeds.	The faced brickwork previously measured as common brickwork is now adjusted. This is a half-brick wall built of facing bricks and finished fair one side. This item of brick walling includes the facing bricks. The description of facework includes pointing, as well as bricks, bond and mortar. The estimator needs to know the type of brick to assess the labours involved.
			22·580 less 4/247·5 990 21·590	Note method of arriving at mean girth of foundation spread outside building, working from length of outside face of foundation trench previously obtained for earthwork support.
	21·59 0·25 0·15	0·81	Fillg. to excavns., av. thickness ≤0·25 m, from on site spoil heaps, topsoil.	Adjustment of topsoil outside building, obtaining the topsoil from on site spoil heaps (SMM D20.9.1.2.3). Plant and protection items are covered in the Preliminaries Bill items.
	Item		Disposal of surf. water.	Item included as SMM D20.8.1.0.0. It is assumed that the excavation work is above normal groundwater level, otherwise an additional item will be required in accordance with SMM D20.8.2.0.0.

1.4

BASEMENT

Timesing	Dimensions	Description
		up to dpc level
		add sprd. of founds. 2/100 — 3910, 3·410; 200, 200; 4·110, 3·610
	4·11 3·61	Exc. topsoil for preservn av. 150dp.
	4·11 3·61 0·15	Disposal of excvtd. mat. on site in spoil heaps av. dist. of 20 m from excavn.
		above grd. 375 less susp. slab 150 225 depth 2·500 less above grd. 225, topsoil 150 — 375 2·125 add upper flr. slab 150, asp. 30, lower flr. slab 100 — 280 2·405
		add fdn. proj. 2/100 — 3·910, 3·410; 200, 200; 4·110, 3·610
	4·11 3·61 2·41	Exc. bast. max. depth ≤ 4·00 m. & Disposal of excavtd. mat. off site.
		3·910 3·410 2/7·320 14·640
		add 2·125 topsoil 150 upper flr. slab. 150 asp. 30 — 330 2·455

2.1

EXAMPLE II

Note addition of 100 mm from outer face of wall to allow for spread of foundations to basement.

Excavating preserved topsoil is measured in m^2 stating the average depth (SMM D20.2.1.1.0).

Disposal of topsoil is measured in accordance with SMM D20.8.3.2.1, assuming that its final distribution will be dealt with in the external works section.

Note the use of calculations in waste for building up the depth of excavation.

The volume of basement excavation is taken to the outside face of the concrete foundation to the basement, projected vertically upwards to the underside of the topsoil. A working space allowance to excavations will subsequently be measured in m^2 adjacent to the half-brick lining wall, but not to the concrete foundations which will be cast against the earth face (See SMM D20.M7–8).

Basement excavation is measured in m^3 stating the starting level, where >0·25 m below existing ground level, and giving the maximum depth range as SMM D20.2.3.4.1.

The full volume of excavation is taken as disposal off site as SMM D20.8.3.1.0 in the first instance, and the soil outside the basement, is later adjusted as filling to excavations (SMM D20.9).

Note calculation of girth of working space in waste, being the girth of the outer face of the protective walls (SMM D20.M8).

The depth is taken from ground level to the top of the basement foundation.

BASEMENT

Drawing No. 2

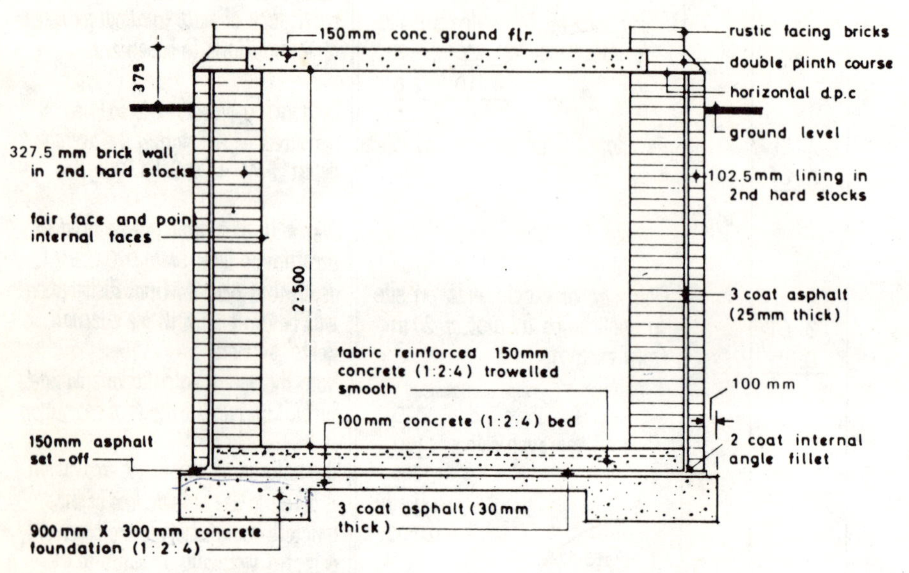

SECTION

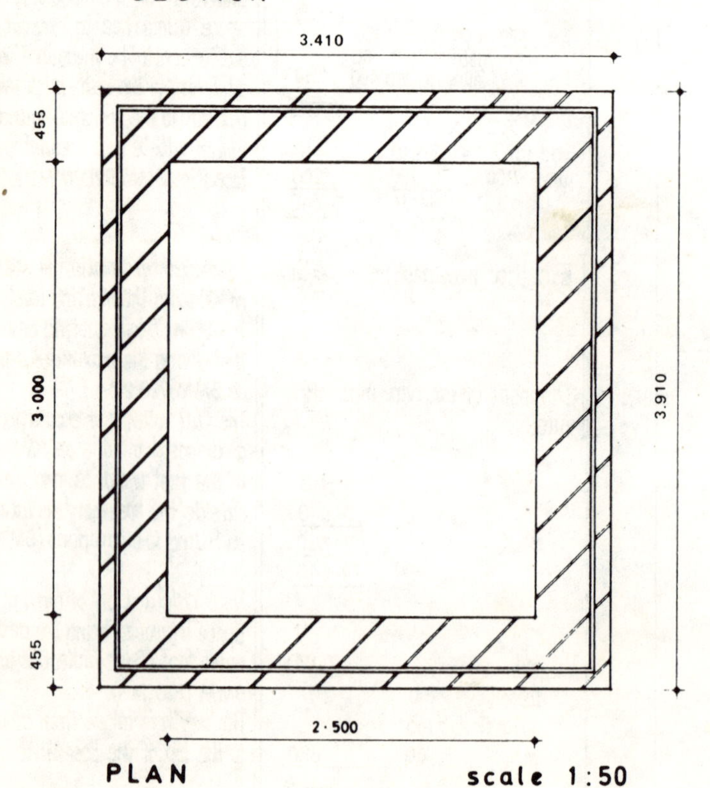

PLAN

scale 1:50

	BASEMENT	(Contd.)	
	14.64 2.46	Wkg. space allce. to excavns. for bast.	Working space allowance to basement excavation is measured in m^2 as a single item in accordance with SMM D20.6.1.0.0. Additional earthwork support, disposal and backfilling are deemed to be included (SMM D20.C2). Every step in the preliminary calculations is carefully set down in 'waste' irrespective of how elementary it may be. All dimensions in 'waste' are taken to the nearest millimetre, unlike those in the dimension column which are entered to the nearest 10 mm.
		Fdn. tr. fdn. depth 300 less bed 100 depth below g.l. 200 2.455 add bed 100 2.555 len. 4.110 3.610 2/ 7.720 15.440 less corners 4/2/450 3.600 11.840	
	11.84 0.90 0.20	Exc. tr., width > 0.30 m, max. depth ≤ 0.25 m, commg. 2.56 m below g.l. & Disposal of excvtd. mat. off site.	Trench excavation is measured in accordance with SMM D20.2.6.1.1. No returned filling is involved in this case.
		Earthwk. suppt. depth 2.455 add fdn. 300 2.755	
2/	4.11 2.76	Earthwk. suppt., max. depth ≤ 4.00 m, dist. between opposg. faces: 2.00 – 4.00 m.	Earthwork support is measured in m^2 and classified by maximum depth stages and giving the range of distances between opposing faces as SMM D20.7.3.2–3.0. It is not measured to the faces of additional excavation resulting from working space (SMM D20.C2). Two separate items are needed for earthwork support to the basement because of the varying distances between the opposing faces.
2/	3.61 2.76	Ditto., max. depth ≤ 4.00 m, dist. between opposg. faces > 4.00 m.	No earthwork support is measured to the inside face of the foundation trench, as it is ≤ 0.25 m in depth (SMM D20.M9a).

2.2

BASEMENT (Contd.)

Dimensions	Description	Side notes
11·84 0·90 0·20	In situ conc. fdns. (1:2:4/40agg) poured on or against earth.	In situ concrete in foundations is measured in m^3 and described as poured on or against earth (SMM E10.1.0.0.5). If the aggregate size had been the same in the bed, this could have been measured as a thickening of the bed (SMM E10. D3).
4·11 3·61	Compactg. bott. of excavn.	Compacting bottom of excavation is measured in m^2 (SMM D20.13.2.3.0).
4·11 3·61 0·10	In situ conc. bed. (1:2:4/20 agg.) thickness ≤ 150, poured partly on or against earth.	Beds of in situ concrete are measured in m^3 giving the thickness classification as SMM E10.4.1–3. Beds poured on or against earth or unblinded concrete are so described (SMM E10.4.1.0.5). It is not however considered necessary to split this particular item.
4·11 3·61	Trowellg. surf. of conc.	Treating the surface of unset concrete is described and measured in m^2 (SMM E41.1–7).
	Asp. tankg. bk. ling. 102·5 proj. conc. 100 202·5 less asp. set-off 150 52·5 4·110 3·610 less 2/52·5 105 105 4·005 3·505	
4·01 3·51	Mastic asp. tankg. & damp prfg. to BS 6577, width > 300, hor., in 3 cts. fin. 30 th. on conc. bed & subseqy. covd.	Asphalt is measured in m^2 under the appropriate classifications and stating the width range and pitch (SMM J20.1.4.1.1) and giving the particulars required by SMM J20.S1–4, and reference to subsequent covering of the asphalt where applicable.
	3·000 2500 add walls 2/327·5 655 655 3·655 3·155	
3·66 3·16 0·15	In situ conc. bed (1:2:4/20 agg.) thickness ≤ 150, reinfd., laid on asp. (m/s).	Where concrete floors are laid over asphalt, this should desirably be stated, because of the possibility of damage to the asphalt and the extra care needed.

2.3

BASEMENT (Contd.)

Dimensions	Description / waste
	3·655 3·155
	less conc. cover 2/50 100 100
	3·555 3·055
3·56 3·06	Stl. fabric reinft. to BS 4483, ref. A193, weighg. 3·02 kg/m², w. 150 min. laps.
3·00 2·50	Trowellg. surf. of conc.
	3·655 3·155 2/6·810 13·620
13·62	Fwk. to edges of beds, plain vert., ht. ≦ 250.
	Asp. L. fillet 3·000 2·500 2/5·500 11·000 add walls 4/2/327·5 2·620 v. asp. 4/2/25 200 13·820
13·82	Asp. int. L. fillet, 40 face.
	Bwk. int. gth. 11·000 add corners 4/327·5 1·310 12·310
12·31 2·50	Bk. wall, facewk. as thickness: 327·5, in 2nd. hd. stocks in English bond in c.m. (1:3).
	12·310 add corners 4/327·5 1·310 13·620

2.4

Steel fabric reinforcement is measured net in m² stating the mesh reference, weight/m² and minimum laps, as SMM E30.4.1.0.0 and E30.S4. Tying wire, cutting, bending, laps and spacers and chairs at the discretion of the contractor are deemed to be included (SMM E30.C2).
Worked finishes to concrete are measured in m² (SMM E41.1–7).

Formwork required to edge of bed as SMM E20.2.1.2:0.
Build-up of girth of asphalt internal angle fillet at junction of horizontal and vertical tanking, measured on the surface of the vertical tanking, as the asphalt is measured the area in contact with the base, making no allowance for the thickness of the asphalt SMM J20.M3.
Angle fillets are measured in metres with a dimensioned description (SMM J20.12.1.0.1) and are deemed to be in two coats, unless otherwise stated, and to include ends and angles. Rough cutting to brickwork to accommodate the fillet is deemed to be included in the brickwork rates (SMM F10.C1c).

Measurement of brickwork up to horizontal dpc, using the internal girth dimension previously calculated in waste.
Brickwork is measured in m² stating the thickness and classified as in walls (SMM F10.1.2.1.0). The facework on one side (internally) is work in bricks finished fair (SMM F10.D2). The description is to include the kind, quality and size of bricks, bond and composition and mix of mortar (SMM F10.S1–3). Brick sizes are best given in a preamble clause.
An alternative method of calculating the outside girth is shown in waste.

BASEMENT (Contd.)

		ht. 2·500 add flr. 150 2·650	Raking out joints of brickwork to form a key for asphalt is deemed to be included in the brickwork (SMM F10.C1d).
13·62 2·65		Mastic asp. tankg. & dampprfg. to BS 6577, width > 300, vert. coverg. in 3cts. fin. 25 th., on bwk. subseqy. covd.	Asphalt is classified in accordance with SMM J20.1.4.1.1 and descriptions include the particulars in SMM J20. S1–4. The asphalt is measured the area in contact with the base (SMM J20.M3).
		len. 13·620 add asp. 4/2/25 200 ling. 4/102·5 410 14·230	
14·23 2·65		Bk. wall, thickness : 102·5, in 2nd hd. stocks in stret. bond in c.m. (1:3), built against asp.	It is necessary to state that the brickwork is to be built against another material, in order that the estimator can allow for the extra cost involved (SMM F10.1.1.1.1).
		ht. 375 less g.f. slab 150 225 add facewk. below g.l. 75 300	
14·23 0·30		Ddt. ditto & Add do., facewk. o.s. in multi-col. stocks basic price £260/1000 (delvd. to site) & flush ptd. as wk. proceeds.	Adjustment of brickwork in facings in place of 2nd class stocks on outer face from dpc down to one course below ground level.
		11·000 add corners 4/4.55 1·820 12·820	The overall thickness of wall is taken from the figured dimension on the drawing.
12·82 0·46		Mastic asp. a.b. in tankg. & dampprfg., width > 300, hor., in 3 cts. fin. 30th. on bwk. subseqy. covd.	This has been measured in accordance with J20.1.4.1.1. Alternatively, it would presumably be permissible to follow the procedure described in F30.2.2.3.0, if it was regarded solely as a dpc.

2.5

BASEMENT (Contd.)

	15·440 less corners 4/2/50 400 15·040 depth 2·755 less fdn. 300 topsoil 150 450 2·305	
15·04 0·10 2·31	Ddt. Disposal of excvtd. mat. off site. & Add Fillg. to excavns., av. thickness > 0·25 m, arising from excavns.	Adjustment for backfilling of soil outside the basement but not including working space. The latter is deemed to be included in the working space item. The filling is measured in accordance with SMM D20.9.2.1.0.
15·04 0·10 0·15	Fillg. to excavns., av. thickness ≤ 0·25 m, obtained from on site spoil heaps, topsoil.	Filling excavations with topsoil outside the building to a depth of 150 mm (SMM D20.9.1.2.3).
Item	Disposal of surf. water.	As SMM D20.8.1.0.0. Surface water is water on the surface of the site and the excavations (SMM D20.D9).

2.6

5 Measurement of Brick and Block Walling

MEASUREMENT OF BRICK AND BLOCK WALLING

Measurement Generally

Brickwork and blockwork are measured in m^2 stating the actual thickness, for example 215 mm, the plane where other than vertical, such as battering or tapering, and whether it has facework (finished fair) on one or both sides (SMM F10.1.1–3.1–4). It must include full particulars of the bricks or blocks, type of bond, composition and mix of mortar and type of pointing (SMM F10.S1–4).

With hollow walls, the skins and the forming of the cavity are each measured separately. The width of the cavity must be stated in the forming cavity item and also the type, size and spacing of wall ties, and the type, thickness and method of fixing any cavity insulation (SMM F30.1.1.1.1 and F30.S2–3. A single comprehensive item of a 255 mm hollow wall including cavity, ties and insulation is not permissible. Walls include skins of hollow walls and so these are not separately described (SMM F10.D4).

The normal order of measurement is: (1) external walls, (2) internal walls and (3) chimney breasts and stacks, as it is advisable to proceed in a logical and orderly sequence.

Various classes of brick and block walling are each kept separate as shown in the following list.

(1) Walling in different bricks, blocks, mortars, bonds or types of pointing (SMM F10.S1–4).
(2) Walls of different thicknesses (SMM F10.1.1).
(3) Walls with facework (finished fair) on one or both sides (SMM F10.1.2–3).
(4) Isolated piers, isolated casings and chimney stacks (SMM F10.2–4).
(5) Battering walls (SMM F10.1.1.2).
(6) Walls tapering one side (SMM F10.1.1.3).
(7) Walls tapering both sides (SMM F10.1.1.4).
(8) Walling used as formwork (SMM F10.1.1.1.3).

(9) Boiler seatings and flue linings (SMM F10.8–9).
(10) Curved work (SMM F10.M4).
(11) Work built overhand (SMM F10.1.1.1.4).
(12) Work built against or bonded to other work (SMM F10.1.1.1.1–2).
(13) Projections (SMM F10.5).
(14) Arches (SMM F10.6).
(15) Isolated chimney shafts (SMM F10.7).
(16) Closing cavities (SMM F10.12).
(17) Bonding to existing (SMM F10.25).
(18) *Facework* ornamental bands, quoins, sills, thresholds, copings, steps, tumblings to buttresses, key blocks, corbels, bases to pilasters, cappings to pilasters and cappings to isolated piers (SMM F10.13–24).

External Walls

The walling above and below the damp-proof course is often measured separately for convenience as a natural demarcation line, and may be built in different bricks or blocks and/or mortars. It must be emphasised that facework is now included in the measured items for brick and block walling. Work is deemed to be vertical unless otherwise described (SMM F10.D3).

The length of external walling will be obtained by the method of girthing illustrated in chapter 3 and the height will normally be taken up to some convenient level, such as the general eaves line. Any additional areas of external wall, such as gables, parapets and walling up to higher eaves levels, will then be taken off. Adjustment of walling for window and door openings will be made when measuring the windows and doors.

The measurement of the areas of external walls will be followed by incidental items, such as projections, facework ornamental bands and facework quoins, normally working from ground level upwards. Facing bricks are generally taken from 75 mm (one course) below ground level, to allow for any irregularities, to just above soffit boarding at eaves.

Walling could either be described as 102.5 or 215 thick or, as in the examples, in this book, as thickness: 102.5 or 215, in a similar manner to that adopted for the measurement of civil engineering work. It is not considered necessary to insert the mm symbols. The preface to SMM7 does permit variations in formulating bill descriptions, including the use of traditional prose. Another alternative would be to describe the walls as half brick or one brick thick. It might be considered that the nominal thickness of a half brick wall as defined in SMM F10.D1 should be 100 mm and not 102.5 mm as used in these examples.

Internal Walls

The measurement of external walls is usually followed by internal walls, which may be of bricks or blocks. A careful check should be made on the type and thickness of each partition, and where there is a number of different types of partition it is often helpful to colour each type in a different colour on the floor plans and suitably to mark each length on the floor plan as it is taken off.

Chimney Breasts and Stacks
This work covers much of the subject matter of chapter 6, which includes a worked example covering a chimney breast and stack.

INCIDENTAL WORKS

Damp-proof Courses
These have already been covered in chapter 4.

Rough and Fair Cutting

All rough and fair cutting is deemed to be included with the brickwork or blockwork (SMM F10.C1b). Similarly, forming rough and fair grooves, throats, mortices, chases, rebates and holes, stops and mitres, and raking out joints to form a key are also deemed to be included (SMM F10.C1 c–d).

Eaves Filling
Brickwork and blockwork in eaves filling is added to the general brick and block walling respectively, and no additional item is required for the extra labour involved (SMM F10.C1e).

Projections

Projections of attached piers, plinths, oversailing courses and the like are measured in metres stating the width and depth of the projection, and whether horizontal, raking or vertical (SMM F10.5.1.1–3.0). Attached pier projections come within this category when their length on plan is ≤ four times their thickness, otherwise they will be measured as walls of the overall thickness (SMM F10.D9).

Figure XVII illustrates the projections to the 440 × 440 mm pier at the end of a 215 mm wall, the projections on each side, being measured in metres,

FIG XVII

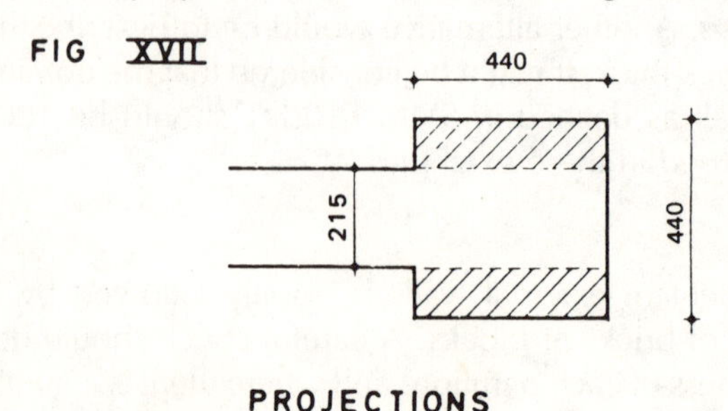

PROJECTIONS

giving the width and depth in the description (440 and 112.5 mm respectively).

Deductions for String Courses and the Like

Deductions for string courses, lintels, sills, plates and the like are measured as regards height to the extent of brick or block courses displaced and as regards depth to the extent only of full half brick beds displaced (SMM F10.M3).

Facework Ornamental Bands

Facework to brick-on-edge bands, brick-on-end bands, basket pattern bands, moulded or splayed plinth cappings, moulded string courses, moulded cornices and the like are each measured separately in metres, giving the width of the band, and where sunk or projecting, the depth of set-back or set-forward, and are normally taken as extra over the work in which they occur (SMM F10.13.1–3.3.1.). Labours in returns, ends and angles are deemed to be included (SMM F10.C1f). The measurement of facework ornamental bands is illustrated in example III.

FACEWORK QUOINS

Facework quoins are formed with facing bricks which differ in kind or size from the general facings (SMM F10.D12). They are measured in metres, stating whether flush, sunk or projecting and giving the appropriate dimensions including the mean girth, and they are normally taken as extra over the work in which they occur (SMM F10.14.1–3.1.1).

WORKED EXAMPLES

Two worked examples follow covering brickwork, facework and block partitions to a small building and a curved brick screen wall.

In all examples in this book, brick dimensions of 215 × 102.5 × 65 mm have been used and 10 mm mortar joints. Thus a half-brick wall is 102.5 mm thick, a one-brick wall 215 mm and a hollow wall with two brick skins 255 mm. Brick sizes will normally be given in a preamble clause at the head of the Brick/Block Walling Bill.

BRICKWORK, BLOCKWORK and FACEWORK to small building

Drawing No. 3

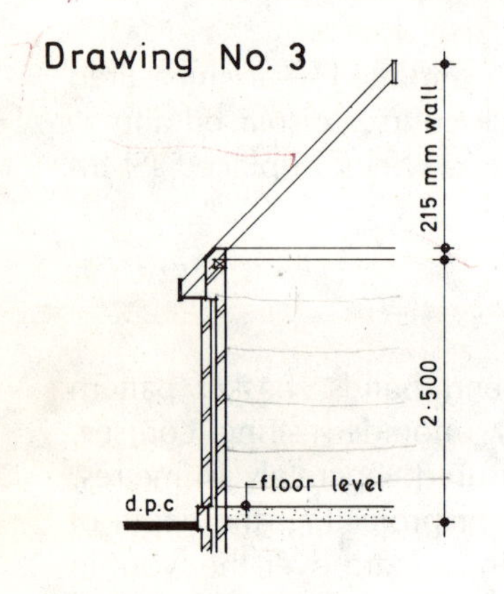

SECTION A–A
scale 1:100

soffit boarding
flush brick on end band
flush band in red facing bricks
facework in multi-coloured bricks
plain band projecting 25 mm (2 courses)
plinth 225 mm wide projecting 40 mm
d.p.c
ground level

ELEVATION scale 1:50

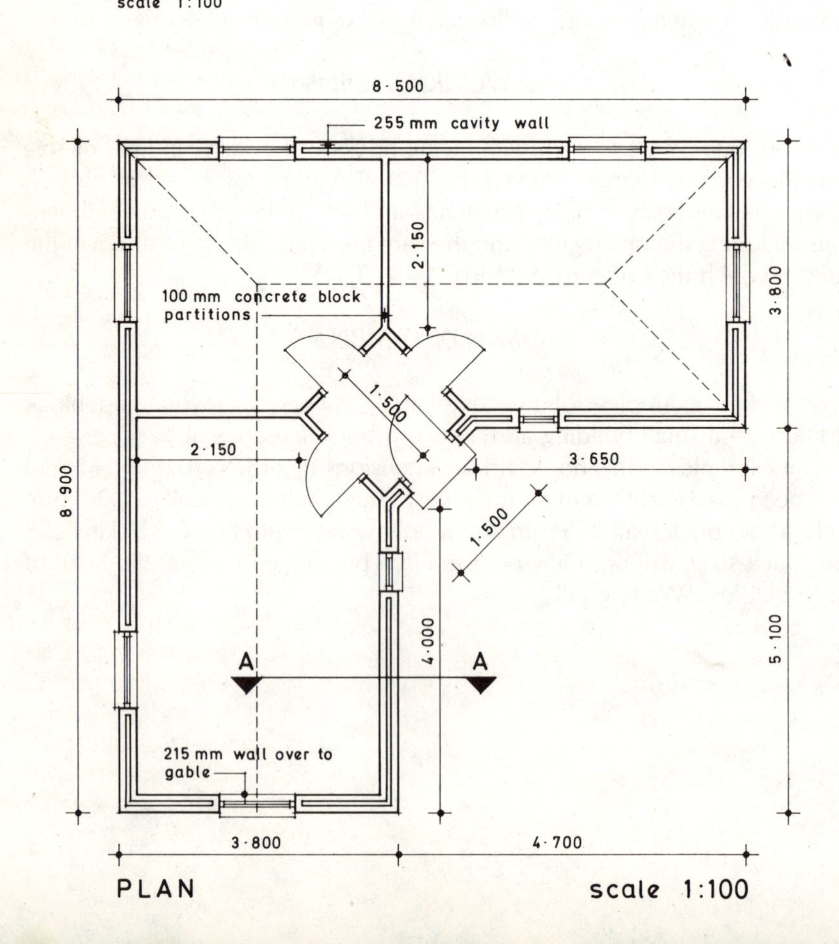

BRICKWORK, FACEWORK AND BLOCKWORK

(brick and block walling above dpc)

(It is assumed that work below dpc has already been measured)

EXAMPLE III

SMM F10.P1 prescribes that drawings accompanying the bill of quantities should show plans of each floor level and principal sections showing the position of and materials used in walls, and external elevations showing the materials used.

External Walls
Length (℄)

	8·500
	8·900
2/	17·400
less	34·800
spld. crnr. 2/1·080	2·160
	32·640
add splay	1·500
	34·140
less crnrs. 4/255	1·020
	33·120
½ thickness of cav.	25
½ thickness of h.b. skin	51·25
	76·25
outer skin	
	33·120
add 4/2/76·25	610
	33·730
inner skin	
	33·120
less 4/2/76·25	610
	32·510

Note the build-up of wall girths in 'waste', including adjustment for the splayed corner.
This is best drawn to a large scale, such as 1:20, or calculated from the dimension on the outer face of the wall.

1·080
1·080
1·500

Adjustment is made for the four external angles only, as the splayed corner cancels out one external angle.
Centre line measurements are obtained for both inner and outer skins, and each of these and the formation of cavity, including ties, are measured separately (see SMM F10.1.1.1.0 and F.30.1.1.1.0).
The hollow wall is measured from dpc to soffit board level, with solid brickwork above. Adjustments for window and door openings are made when measuring the windows and doors, and will include closing cavity items as SMM F10.12.1.1.0.0.

32·51 2·10		Bk. wall, thickness: 102·5, comms. in stret. bond in g.m. (1:1:6).

The half brick skins to the hollow wall are measured in accordance with SMM F10.1.1.1.0 and the descriptions are to include the size, kind and quality of bricks, bond, composition and mix of mortar and type of pointing, where appropriate (SMM F10.S1–4).

3.1

BRICKWORK, FACEWORK AND BLOCKWORK (Contd.)

Timesing	Dimensions	Squaring	Description	Side notes
	33·73 2·10		Bk. wall, facewk. o.s., thickness: 102·5, multi-col. fcg. bks. type A, basic price £250/1000 (delvd. to site) in stret. bond in g.m. (1:1:6) & ptg. w. nt. struck jt. as wk. proceeds.	Ideally the contractor should be given precise details of the type of brick to be used so that he can assess the labour involved in laying and cutting the bricks. In practice bricks are often covered by a prime cost sum or basic price per 1000 bricks to keep the choice of brick open and yet ensure that all contractors are tendering on the same basis.
	33·12 2·10		Form cav. in holl. wall, width: 50, inc. 4nr. wall ties/m² of zinc coated mild steel vert. twist type to BS. 1243.	The work connected with the cavity is measured as SMM F30.1.1.1.0. Alternatively the wall tie description could be given in a preamble clause at the head of the brickwork and blockwork section of the Bill, or there could be a cross reference to the project specification. The width of the cavity behind the projecting plinth would be increased to 90 mm and would thus form a separate cavity item.
			1B wall len. 33·120 less crnrs. 2/40 80 33·040 less gable 3·800 29·240	Fair birdsmouth angles are deemed to be included (SMM F10.C1f). The centre lines of the 215 and 255 walls do not coincide, and hence a new girth dimension has to be calculated for the one brick wall.
	29·24 0·50		Bk. wall, thickness: 215, comms. in English bond in g.m. (1:1:6).	Solid unfaced brickwork at head of walls is measured to underside of roof coverings, excluding the gable where the cavity wall is carried up to a higher level (top of ceiling joists).
	3·80 0·50		Bk. wall, thickness: 102·5, comms. a.b.	Labour in eaves filling is deemed to be included (SMM F10.C1e).
			&	
			Ditto. facewk o.s., multi-col. fcg. bks, type A, a.b. (gable up to clg. lev.	Higher section of hollow wall at gable from eaves up to top of ceiling joists.
			&	Note the use of 'a.b.' to avoid repeating a description which has already been given.
			Form cav. to holl. wall, width: 50, a.b. 3.2	

BRICKWORK, FACEWORK AND BLOCKWORK (Contd.)

Timesing	Dimensions	Squaring	Description	Notes
2/	0.50		Close 50 cav. vert. wall w. 102.5 bwk. & bit. felt based dpc.	Closing cavities at ends of hollow walls or jambs of openings is measured in metres, stating the width of cavity and method of closing (SMM F10.12.1.1.0). The vertical damp-proof course has been included in the description – alternatively it could form a separate item in accordance with SMM F30.2.1.1.0.
			less ends covd. by roof slope 2/150 — 3.800, 300, 3.500	
½/	3.50 1.75		Bk. wall, facewk. o.s., thickness: 215, multi-col. fcg. bks. type A in Flem. bond in g.m. (1:1:6), & ptg. w. struck. jt. as wk. proceeds. (gable	Area of triangular section of brickwork to gable = ½ height x base.
			Plinth: 34.140; add crnrs. 4/2/40 — 320; 34.460	The plinth is below dpc but has been included in the dimensions to illustrate the method of measurement in accordance with SMM F10.13.3.3.1. The plinth is measured in metres giving the description and dimensions prescribed in SMM F10.13.3.3.1 and F10.D11.
	34.46		Facewk, ornamental bk-on-end band projectg., depth of set fwd. 40, hor., width: 225, e.o. wk. in which it occurs, multi-col. fcg. bks., type A, in c.m. (1:3) & ptg. w. flush jt. as wk. proceeds.	No adjustment is made for the areas of faced brickwork occupied by the facework ornamental band, as it has been taken as extra over the work in which it occurs. Labours in ends and angles are deemed to be included (SMM F10.C1f). It would also be necessary to adjust for the wider cavity against the plinth in a similar manner to the later item in connection with the upper projecting band.
			Projctg. courses len.: 34.140; add crnrs. 4/2/25 — 200; 34.340	
	34.34		Proj. in facewk., width: 150, depth of proj. 25, hor. in multi-col. fcg. bks. a.b.	These two projecting courses have been measured in accordance with SMM F10.5.1.3.0. as projections, as they do not appear to meet the requirements of facework ornamental bands as defined in SMM F10.D11. No adjustment of brickwork previously measured is required as it only entails additional labour which is covered by the measured item.
			len.: 33.120; add crnrs. 4/2/12.5 — 100; 33.220	Build-up of the slightly increased length of wider cavity.

3.3

BRICKWORK, FACEWORK AND BLOCKWORK (Contd)

Dimensions	Description	Notes
33·22 0·15	Form cav. in holl. wall, width: 75 inc. wall ties a.b.	Cavities of differing widths must be kept separate (SMM F30.1.1.1.0).
33·12 0·15	Ddt Form cav. in holl. wall, width: 50.	
	Flush bands	A description of the different types of brick is required (SMM F10.S1). Both bands are measured as facework ornamental bands in accordance with SMM F10.13.1.3.1.
34·14	Facewk. ornamental band, flush, hor., width: 150, e.o. wk. in which it occurs, in stretcher bond, red fcg. bks. type B, basic price £270 per 1000 (delvd. to site), in g.m. (1:1:6) & ptg. w. wt. struck jt. as wk. proceeds. & Facewk. ornamental bk-on-end band flush, hor., width: 225, e.o. wk. in which it occurs, multi-col. fcg. bks. type A in g.m. (1:1:6) & ptg. w. flush jt as wk. proceeds.	Both bands have the same length, assuming both continue across the gable wall, and so they are grouped together. No adjustment is made for the areas of faced brickwork occupied by the facework ornamental bands, as they have been taken as extra over the work in which they occur.
	Intl. Ptns. len. 2/2·150 4·300 2/1·500 3·000 1·700 9·000	Note build-up of total length of partitions. A full description of the blocks is given in accordance with SMM F30.S1. Alternatively this information could be given in a preamble or a cross reference made to the project specification.
9·00 2·50	Conc. blk. wall, thickness: 100, type B lightwt. agg. blks, size 440 x 215 to BS 2028, w. keyed fin., b.&j. in g.m. (1:1:6).	No pointing is required as it is assumed that the internal wall surfaces will be plastered. Adjustments for door openings will be taken when measuring the doors. No additional item has been taken for the bonding of the ends of the block walls to brickwork, as forming chases is deemed to be included under SMM F10.C1c, and the measurement of bonding to other work of differing material, referred to in SMM F10.M5, relates only to superficial items of work, and its inclusion can be inferred from SMM F10.C2.

3.4

CURVED BRICK SCREEN WALL

EXAMPLE IV

Timesing	Dimensions	Description	Notes
		<u>Work below dpc</u> diam. int. rad. 2·500 ½/215 107·5 2/ 2·607·5 mean diam. 5·215	The measurement of the foundations has been included in this example, as various points of value to the student emanate from the curved construction.
¼/22/7/	5·22 0·52	Exc. topsoil for preservn. av. 150 dp. (wall	Topsoil excavation is measured in m² as SMM D20.2.1.1.0, although the more usual approach in this type of situation would probably be to commence trench excavation from ground level.
2/	0·63 0·63	(piers	Circumference of quadrant (¼ circle) = ¼ π D. There is a higher cost involved in excavating topsoil from an area bounded by curved sides and it might be argued that SMM General Rules 1.1 could be invoked, ie. more information than is required by the SMM shall be provided where necessary. However, the curved outline will be evident from the drawings supplied with the bill of quantities, and there is no provision in SMM D20 for curved work to be kept separate as there was in SMM 6.
		628 328 2)300 150	
2/	0·15 0·52	<u>Ddt</u>. exc. topsoil a.b. (junctn. of wall & pier fdns.	An adjustment is necessary because of the overlapping of excavation measured at junction of wall and pier foundations.
¼/22/7/	5·22 0·52 0·15	Disposal of excvtd. mat. on site in spoil heaps av. dist. of 20 m from excavn.	Separate cubic disposal on site item as SMM D20.8.3.2.1.
2/	0·63 0·63 0·15		
2/	0·15 0·52 0·15	<u>Ddt</u>. do. (junctn. of wall & pier fdns.	An adjustment item covering the overlap at the junction of wall and pier foundations.
			The SMM does not require excavating curved trenches to be kept separate from straight trenches. SMM D20.C3 further prescribes that curved earthwork support is deemed to include any extra costs of curved excavation.

Drawing No. 4

CURVED BRICK SCREEN WALL

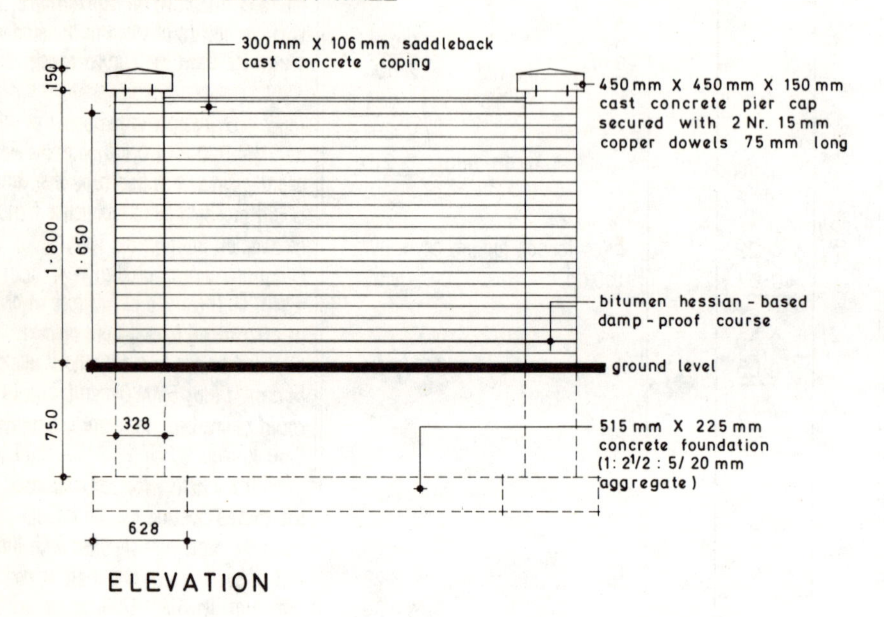

ELEVATION

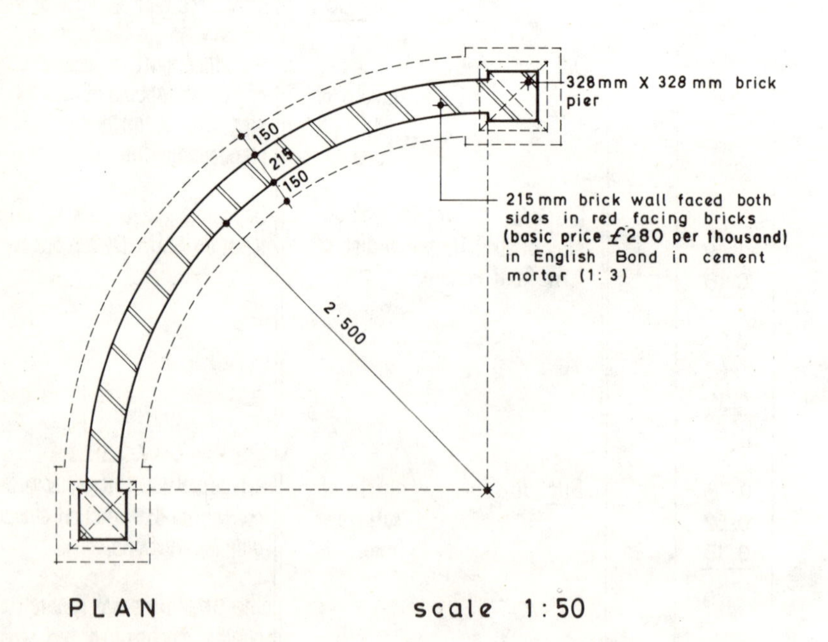

PLAN scale 1:50

CURVED BRICK SCREEN WALL (Contd.)

Timesing	Dimensions	Squaring	Description	Side notes
			depth 750 add fdn. 225 975 less topsoil 150 825	The excavation is measured in m^3 in the maximum depth ranges listed in SMM D20.2.6.2.0. The trench excavation and filling includes that to the pier foundations. The commencing level of the excavation is not stated unless it exceeds 0·25 m below existing ground.
$\frac{1}{4}$/$\frac{22}{7}$/	5·22 0·52 0·83		Exc. tr. width > 0·30 m, max. depth ≤ 1·00 m. (wall &	
2/	0·63 0·63 0·83		Fillg. to excavns., av. thickness > 0·25 m, arisg. from excavns. (piers	Filling returned to trenches is measured in m^3 and classified as SMM D20.9.2.1.0.
2/	0·15 0·52 0·83		Ddt. last two items. (junctn. of wall & pier fdns.	Adjustment of overlap of wall and pier foundations.
$\frac{1}{4}$/$\frac{22}{7}$/	5·22 0·52		Compactg. bott. of excavn. (wall	Compacting the bottoms of excavations in accordance with SMM D20.13.2.3.0.
2/	0·63 0·63		(piers	
2/	0·15 0·52		Ddt. do. (junctn. of wall & pier fdns.	Adjustment of overlap at junction of wall and piers.
2/$\frac{1}{4}$/$\frac{22}{7}$/	5·22 0·98		Earthwk. suppt., max. depth ≤ 1·00 m, dist. between opposg. faces ≤ 2·00 m, curved.	Earthwork support is measured in m^2 and classified in maximum depth ranges as SMM D20.7.1–3. Curved earthwork support is so described (SMM D20.7.1.1.1) irrespective of the radius.
2/2/	0·15 0·98		Ddt. do. (junctn. of wall & pier fdns.	Adjustment needed at junction of wall and pier foundations.

4.2

CURVED BRICK SCREEN WALL (Contd.)

Timesing	Dimensions	Squaring	Description	Notes
			628 515 2) 113 56·5	
2/3/	0·63 0·98		Earthwk. suppt. max. depth ≤ 1·00 m, dist. between opposg. faces ≤ 2·00 m. (sides to piers	Earthwork support measured around pier excavations – 3 sides and 2 returns to each pier.
2/2/	0·06 0·98		(retns. to piers	
1/4/22/7/	5·22 0·52 0·23		In situ conc. fdns. (1:2½:5/20agg.) poured on or against earth. &	In situ concrete foundations are measured in accordance with SMM E10.1.0.0.5. Note method of adjustment of filling for volume occupied by concrete.
2/	0·63 0·63 0·23		Ddt. Fillg. to excavns. a.b.d. & Add Disposal of excvtd. mat. off site.	
2/	0·15 0·52 0·23		Ddt. In situ conc. fdns. (1:2½:5/20agg.) a.b. (junctn. of wall & pier fdns & Add Fillg. to excvns. a.b.d. & Ddt. Disposal of excvtd. mat. off site.	Note method of adjusting filling and disposal of excavated material off site (the reversal of the previous procedure). Walls are measured in m², giving the actual thickness of the wall (SMM F10.1.1.1.0). Walling below dpc is often in a different mortar to that above.
			Bwk. in fdns. ht. 750 g.l. to dpc 150 900	
1/4/22/7/	5·22 0·90		Bk. wall, thickness: 215, curved to 2·61 m rad., comms. in English bond in c.m. (1:3).	The radius is to be stated in the description of curved brickwork (SMM F10.M4) and SMM General Rules 14.2 prescribes that the radius is the mean radius unless otherwise stated. Also include the kind, quality and size of bricks, bond, composition and mix of mortar (SMM F10.S1–3).

CURVED BRICK SCREEN WALL (Contd.)

1/4/22/7/	5·22 0·23		Ddt Bk. wall, thickness : 215, curved to 2·61 m rad., commons, a.b. & Add Bk. wall, thickness : 215, facewk. b.s., curved to 2·61 m rad., red fcg. bks. type C, basic price £280 per thsd. (delivered to site) in English bond & ptg. in c.m. (1:3) w. nt. flush jt. as the wk. proceeds.	The size of bricks and other supporting information could however be included in a preamble clause or by cross reference to the project specification where applicable. Facings are normally taken to one course below ground level, to allow for possible irregularities in the final ground surface. Hence an adjustment is needed to cater for the facework up to damp course level.
2/	0·33 0·90		Bk. wall thickness : 215, comms. in English bond in c.m. (1:3). (piers	The wall is measured through the pier, and the projections beyond the wall face in attached piers are measured in metres, stating the width and depth of the projection (SMM F10.5.1.1.0). Adjustment of piers in facings for those in commons. Note method of calculating depth of brick projections to piers in waste.
2/	0·33 0·23		Ddt. Ditto & Add Do, red fcg. bks. type C in Eng. bond in c.m. (1:3) (piers 328 215 2)113 56·5	
2/2/	0·90		Projs. width : 328, depth : 57, comms. in English bond in c.m. (1:3). (piers	Adjustment of faced projections for plain projections. No additional facework is measured to the ends of the piers as the facing bricks have already been taken and the additional labour is deemed to have been included (SMM F10.C1f).
2/2/	0·23		Ddt Ditto & Add Do, in facewk. o.s., red fcg. bks. type C, a.b. (piers	
1/4/22/7/	5·22 0·22		Dpc, width ≤ 225, hor., curved, single layer of hessian - based bit. felt to BS 743 ref. A, & bedded in c.m. (1:3). (wall	Note description of curved dpc in accordance with SMM F30.2.1.3.0 and F30.M1. It is not necessary to state the radius. The description is to include the gauge, thickness or substance of sheet material, number of layers, and composition and mix of bedding material measured in m², and distinguishing between those ≤ and > 225 mm in width.

4.4

CURVED BRICK SCREEN WALL (Contd.)

Timesing	Dimensions	Squaring	Description	Side notes
2/	0.33 0.33		Dpc, width > 225, hor., single layer of hessian-based bit. felt a.b. (piers	
			Adjust. of excavn. depth of bwk. below grd. 750 less topsoil 150 600	
1/4/22/7/	5.22 0.22 0.60		Ddt. Fillg. to excvns. a.b.d. (wall &	Adjustment of soil disposal for volume of brickwork in the depth of the trench, below topsoil.
2/	0.33 0.33 0.60		Add Disposal of excvtd. mat. off site. (piers	
			2/628 1.256 2/328 656 1.912 less 215 2/150 300 515 1.397	Note build-up of girth of topsoil adjustment around the two piers.
2/1/4/22/7/	5.22 0.15 0.15		Fillg. to excavns., av. thickness ≤ 0.25 m, obtained from on site spoil heaps, topsoil. (wall	Adjustment of topsoil around the outside of the wall and piers, measured in m³ in accordance with SMM D20.9.1.2.3.
2/	1.40 0.15 0.15		(piers	
	Item		Disposal of surf. water.	See SMM D20.8.1.0.0.
			Work above dpc. wall ht. 1.650 less dpc to g.l. 150 1.500	
1/4/22/7/	5.22 1.50		Bk. wall, thickness: 215, facewk. b.s., curved to 2.61 m rad., red fcg. bks., type C, in English bond & ptg. in c.m. (1:3) a.b.d.	Note use of letters a.b.d. (as before described) to abbreviate a description where an identical item has been taken off previously. A wall with facework on both sides must inevitably include pointing on both faces without being specifically mentioned.

4.5

CURVED BRICK SCREEN WALL (Contd.)

		pier ht. 1.800 less dpc to g.l. 150 1.650	
2/	0.33 1.65	Bk. wall, thickness: 215, red. fcg. bks. type C in Eng. bond in c.m. (1:3). (piers	Measurement of piers above dpc level – using the same procedure as before – one-brick wall and projections, all in facing bricks.
2/2/	1.65	Projs., width 328, depth: 57, facewk. o.s., red fcg. bks. type C, in Eng. bond & ptg. in c.m. (1:3) a.b.d. (piers	Projections measured in accordance with SMM F10.5.1.1.0, with the inclusion of facework.
		Copg.	
1/4/22/7/	5.22	Precast conc. copg., saddleback, 300 x 106 to BS 3798, type 1, curved to rad. of 2.61m & b.&p. in c.m. (1:3). (In 5nr. units).	Reference to BS 3798, type 1, avoids the need for including twice weathered and twice throated in the description. Linear item in accordance with SMM F31.1.2.0.0, and including the number of units in the description.
2/	1	Precast conc. pier cap, 450 x 450 x 150, 4 times wethd. & thro. a/r. to match copg., inc. b.&p. in c.m. (1:3), fxd. w. cop. dowel, 15 dia. & 75 lg.	Pier caps are enumerated with a dimensioned description as SMM F31.1.1.0.0, and including method of bedding and fixing (SMM F31.S4).

6 Measurement of Chimney Breasts and Stacks, Fireplaces, Vents and Rubble Walling

CHIMNEY BREASTS AND STACKS

Brickwork in Breasts and Stacks

The projecting chimney breasts and chimney stacks are not usually measured when the general brickwork to a building is being taken off. Where a chimney breast is located on an external wall, the chances are that the wall at the back of the fireplace with its external facework will be measured with the general brickwork, and the projecting breasts and chimney stack will be left to be measured later. With fireplaces on internal walls, it is probable that the whole of the enclosing brickwork will be measured together following the taking off of the general brickwork.

The brickwork projecting from the face of the wall on which the chimney breast is located, will be measured as projections in accordance with SMM F10.5.1.1.0 in metres, stating the width and depth of projection, where the length on plan does not exceed four times the thickness, otherwise it will be measured as a wall of the combined thickness as illustrated in example V. Brickwork in chimney stacks is measured separately in m^2 (SMM F10.4.1–3.1.0).

No brickwork will be taken for the void occupied by the fireplace opening, but the brickwork will be measured solid above the fireplace opening where the extra labour in 'gathering over' is offset by the saving in brickwork.

Flues

In the measurement of chimney flues, no deduction of brickwork for voids will be made when the void is $\leq 0.25\ m^2$ in cross-sectional area (SMM F10.M2b). The normal domestic flue measures 225 × 225 mm ($0.051\ m^2$), so that it would have to be a very large flue for the deduction provision to operate. However, when dealing with a stack containing several flues, their combined total area has to be assessed against the $0.25\ m^2$ cross-sectional area requirement.

Clay and precast concrete flue linings are measured in metres, giving a dimensioned description, and with cutting to form easings and bends and cutting to walls around linings deemed to be included (SMM F30.11.1.0.0 and F30.C7). Brick flue linings are measured in square metres stating the thickness in accordance with SMM F10.9.1.0.0).

Chimney pots are best enumerated, with a dimensioned description and manufacturer's reference, under the classification of proprietary items (SMM F30.16.1.1.0).

FIREPLACES

The general approach to the measurement of fireplaces is illustrated in worked example V, although SMM7 gives no guidance on the measurement of stoves and surrounds, and hence they would come under the classification of proprietary items (SMM F30.16.1.1.0). The supply of stoves or grates and slabbed tile surrounds and hearths is frequently covered by prime cost sums or basic prices, but the work in fixing and supply of incidental materials has also to be taken.

Consideration should also be given at this stage to any air ducts laid under solid floors and connected to fireplaces. The ducts would normally be measured in metres with a full description and any fittings enumerated.

VENTS

Vents are often required to provide ventilation under hollow boarded floors and to toilets and larders. The formation of the opening and the building in of the air brick or ventilating grating are deemed to be included in the enumerated component item. Air bricks, ventilating gratings and soot doors are enumerated, giving the size of opening and nature and thickness of the wall and including lintels and arches where required (SMM F30.12–14.1.0.1–2). Care must be taken not to miss the provision of ventilating gratings, often of metal or fibrous plaster, on the inside face of the wall. Working wall plaster around them is deemed to be included in the wall plaster items.

The following example illustrates the method of measuring a typical vent.

				Notes
	1		215 × 215 precast conc. air bk. to BS 493 in bk. wall 102.5 th. & 215 × 215 gtg. of aluminium-silicon alloy to BS 493, class 2 louvre design in bk. wall 102.5 th.	Provision of air brick on outside face of wall. Note use of British Standards in descriptions. Provision of ventilating grating on inside face of wall (inner skin of hollow wall).

RUBBLE WALLING

There is a possibility that the measurement of rubble walling may be included in some Part I examination syllabuses in measurement, and so a brief summary is given of the main provisions of the *Standard Method* relating to this section.

Rubble work is generally defined as natural stones either irregular in shape or roughly dressed and laid dry or in mortar with comparatively thick joints. Dry rubble walling is frequently used in boundary walls.

The measurement of this class of work follows the rules prescribed in SMM work section F20 (Natural stone rubble walling), with work measured in m^2 giving the thickness, plane (other than vertical) and any facework. Full particulars of the stone, type of walling, mortar and pointing are to be given (SMM F20.S1–10). Rough and fair square cutting are deemed to be included (SMM F20.C1l), while rough or fair raking or circular cutting are measured in metres stating the thickness (SMM F20.26–27.1.0.0). Levelling uncoursed rubble work for damp-proof courses, copings and the like, labours in returns, ends and angles and dressed margins to rubble walling are deemed to be included (SMM F20.C1g–j).

WORKED EXAMPLES

Worked examples covering a chimney breast, stack and fireplace, and a random rubble boundary wall follow.

CHIMNEY BREAST, STACK AND FIREPLACE — EXAMPLE V

(It is assumed that general excavation, foundations, brickwork and facework to external walls have already been measured)

	Dimensions	Description	Notes
		Chy. breast wk. up to dpc fdn. proj. 940 215 338 553 2)387 193.5 chy. breast len. 2/440 880 665 1.545 add sprd. of fdns. 2/193.5 387 1.932	Additional trench excavation to chimney breast over and above that already measured for the external wall (additional width of 338 mm), on the assumption that reduced level excavation has previously been taken down to underside of hardcore. The depth of the foundation trench is scaled from the drawing. This is regarded as an extension to the normal foundation trench and is therefore billed under the same heading (both items exceed 0.30 m in width).
	1.93 0.34 0.68	Exc. tr. width > 0.30m, max. depth ≤ 1.00 m, commg. from reduced level. & Fillg. to excvns. av. thickness > 0.25 m, arising from excavns.	The whole area of the building is assumed to already have been measured for compacting and no additional area is therefore involved. The commencing level for excavation is given as it exceeds 0.25 m below existing ground level (SMM D20.2.6.2.1).
2/	0.34 0.68	Earthwk. support, max. depth ≤ 1.00 m, dist. between opposing faces ≤ 2.00m. (retns.	Additional earthwork support is required to each end of the chimney breast excavation.
	1.93 0.34 0.30	In situ conc. fdns. (1:3:6/40 agg.) poured on or against earth. & Ddt Fillg. to excvns. a.b.d. & Add Disposal of excvtd. mat. off site.	Additional concrete in foundations to chimney breast, with excavated soil disposal adjustments. In situ concrete foundations are measured in accordance with SMM E10.1.0.0.5.
		ht. 900 add g.l. to dpc 150 1.050	Brickwork in chimney breasts is measured in metres in projections, stating the width and depth of projection, as the length is ≤ 4 times the thickness (SMM F10.5.1.1.0. and F10.D9).
2/	0.44 1.05	Bk. proj., width : 440, depth of proj. 338, comms. in English bond in c.m. (1:3).	Brickwork in foundations is kept separate from that above dpc, because of the different mortars used.

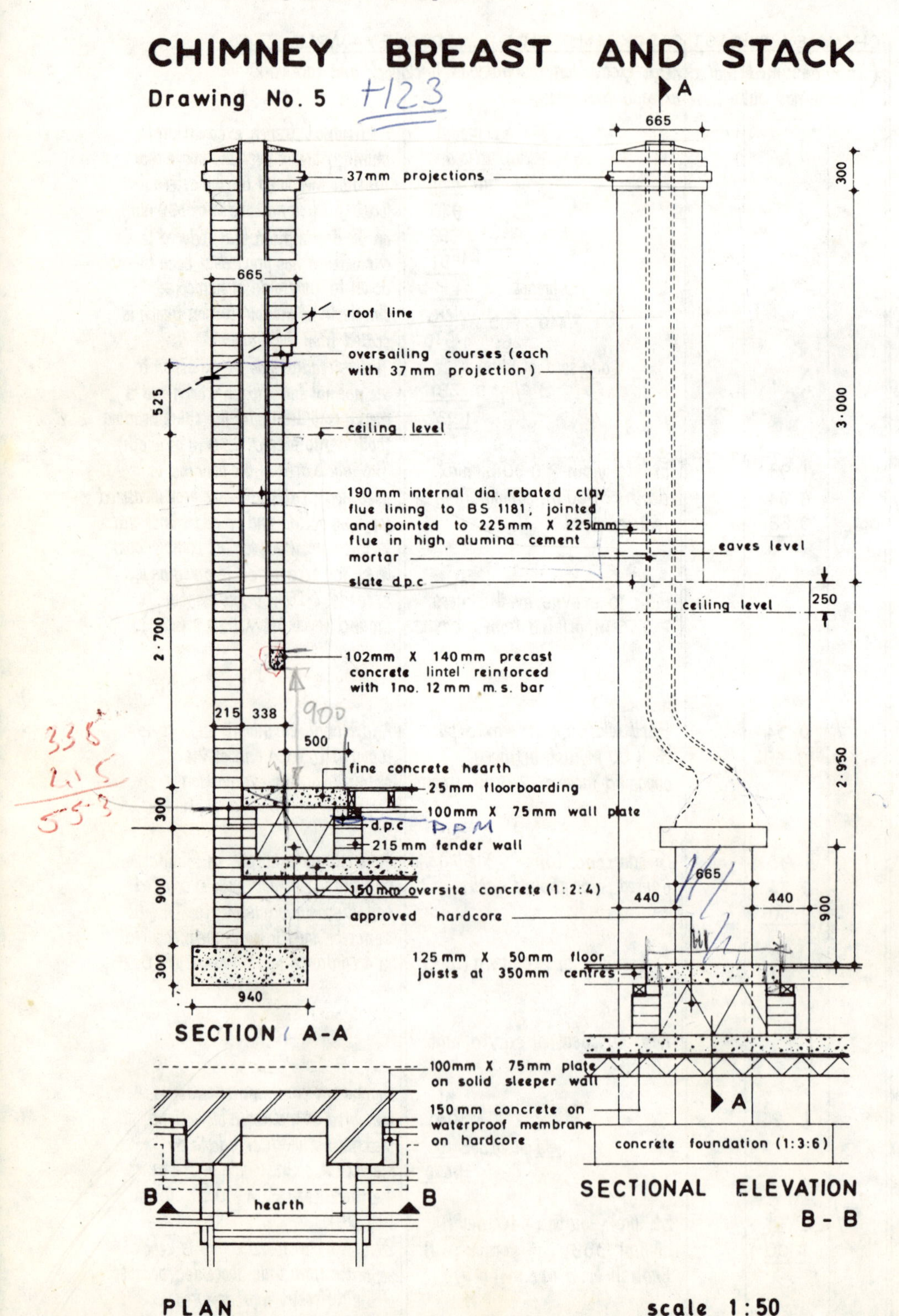
CHIMNEY BREAST AND STACK
Drawing No. 5
A
665
37mm projections
300
665
roof line
oversailing courses (each
with 37mm projection)
525
3·000
ceiling level
190mm internal dia rebated clay
flue lining to BS 1181; jointed
and pointed to 225mm X 225mm
flue in high alumina cement
mortar
eaves level
slate d.p.c
ceiling level
250
2·700
102mm X 140mm precast
concrete lintel reinforced
with 1no. 12mm m.s. bar
215 338
500
fine concrete hearth
25mm floorboarding
2·950
300
100mm X 75mm wall plate
d.p.c
215mm fender wall
665
900
150mm oversite concrete (1:2:4)
440 440
approved hardcore
900
125mm X 50mm floor
joists at 350mm centres
300
940
SECTION A-A
100mm X 75mm plate
on solid sleeper wall
A
150mm concrete on
waterproof membrane
on hardcore
concrete foundation (1:3:6)
B
hearth
B
SECTIONAL ELEVATION
B - B
PLAN
scale 1:50

CHIMNEY BREAST, STACK AND FIREPLACE (Contd.)

Timesing	Dimensions	Description	Notes
		675 300 375	
2/	0.44 0.34 0.38	Ddt. Fillg. to excvns. a.b.d. & Add Disposal of excvtd. mat. off site.	Volume occupied by brickwork below hardcore level. Disposal of excavated material off site measured in accordance with SMM D20.8.3.1.0.
		338 215 553	An adjustment is required to the damp-proof course as it has already been measured for the thickness of the 215 mm wall in accordance with SMM F30.2.1.3.0, as ≤ 225 mm wide, and the combined area of wall and projection now needs to be taken, which places it in the second category (>225 mm).
2/	0.44 0.55	D.p.c., width >225, hor., 2 cos. of stout slates laid bkg. jt. & b.&j. in c.m. (1:3).	
2/	0.44 0.22	Ddt. do. width ≤ 225, do.	
		ht. hth 150 firep. opg. 900 lintel 150 1.200	
		Bwk. above dpc	Gauged mortar has been taken above damp-proof course level.
2/	1.20	Bk. proj., width: 440, depth of proj. 338, comms. in English bond in g.m. (1:1:6).	Projections extend to top of fireplace opening.
		525 2.950 250 1.200 275 1.750	As the length on plan of the brick projection above the fireplace opening exceeds four times the thickness, it cannot be measured as a projection, and is therefore measured as a wall of the combined thickness (553 mm). See SMM F10.D9.
	1.55 1.75 0.67 0.28	Bk. wall, thickness: 553, facewk. o.s. comms. & red fcg. bks. type C in Eng. bond in g.m. (1:1:6) ptd. w. nt. flush jt. as wk. proceeds. (remainder of chy. breast) (above breast) & Ddt. Bk. wall thickness: 215, facewk o.s., red fcg. bks. type C in Eng. bond in g.m. (1:1:6), ptd. a.b.	Adjustment for brickwork already measured. It is assumed that the external wall will have been measured to approximately 525 mm above ceiling level (275 mm above top of chimney breast).

5.2

CHIMNEY BREAST, STACK AND FIREPLACE (contd.)

		Description	Notes
		Chy. stack	Brickwork in chimney stacks is measured separately to walls (SMM F10.4). It is desirable for chimney stacks to be built in cement mortar because of their exposed situations.
	0·67 0·15	Bk. chy. stack, thickness: 553, facewk. o.s., comms. & red fcg. bks. type C, in Eng. bond in c.m. (1:3) ptd. w. nt. flush jt. as work proceeds.	The intention of the Standard Method seems to be that the full cross-section area of the stack should be measured for the damp-proof course, without deduction for the void resulting from the flue, as it does not exceed 0·50 m^2 (SMM F30.M3).
	0·67 0·55	D.p.c., width > 225, hor. a.b.d.	
3/	0·67	Bk. proj., width: 75, depth of proj. 37, hor., comms. in c.m. (1:3). (oslg. cos	Oversailing courses are measured as projections giving the width and depth of the projection (SMM F10.5.1.3.0).
		ht. 275 / 150 / 425 — 3·000 300 3·300 less part already taken 425 2·875	Chimney stack measured up to the top, and projection items for oversailing courses will be taken later. Chimney stack measured in accordance with SMM F10.4.3.1.0.
	0·67 2·88	Bk. chy. stack, thickness: 665, facewk. b.s., red fcg. bks. type C in Eng. bond in c.m. (1:3), ptd. w. nt. flush jt. as wk. proceeds.	The description includes facework (finished fair) to both sides and the extra labour to other two sides is deemed to be included (SMM F10.C1f).
		4/665 2·660 add corners 4/2/37 296 inner cos. 2·956 add 4/2/37 296 outer cos. 3·252	Calculations in waste determine the girth of the outer face of the projecting courses.
2/	2·96 3·26	Bk. proj., facewk., width: 75, depth of proj. 37, hor.	The projections in facework are measured in metres stating the width and depth of the projection (SMM F10.5.1.3.0). Angles are deemed to be included (SMM F10.C1f).
	1	Clay flue terminal to BS 1181, type 6F, 300 lg. & 185 int. diam., proj. 75 above top of chy. stack & flaunchg. in c.m. (1:3).	Terminal is enumerated and described as a proprietary item in accordance with SMM F30.16.1.1.0, even although it is covered by a British Standard as there is no other relevant SMM item. This particular British Standard covers several types of flue terminal and it must be made clear in the description which type is required.

5.3

CHIMNEY BREAST, STACK AND FIREPLACE (Contd.)

Timesing	Dimensions	Squaring	Description	Side notes
			add beargs. 665 2/100 200 865	Build-up of length of lintel over fireplace opening.
	1		Precast conc. lintel (1:2:4/20 agg.) spld. 102 x 140 x 865 lg. reinfd. w. 1 nr. 12ϕ m.s. bar to BS4449 & b.&j. in g.m. (1:1:6).	Lintels are enumerated with a dimensioned description, and giving materials, mix and reinforcement details (SMM F31.1.1.0.1).
			4/665 = 2·660	Moulds are deemed to be included (SMM F31.C1).
	2·66 0·60		Rendered coatg. to bk. walls, width > 300, in isolated areas, in 1 ct. 15th., ct. & sd. (1:3), trowelled.	Rendering to outside face of chimney stack where it passes through roof space. The particulars are given in accordance with SMM M20.1.1.1.0 and M20.S1-6. The work is deemed to be internal unless described as external. It has also been described as work in isolated areas, giving more information than required by the rules, as SMM General Rules 1.1, to assist the estimator in pricing. Rounded angles are measured in metres if in the range 10-100 mm radius (SMM M20.16 and M20.M7). No item has been taken as the radius is likely to be less than 10 mm.
			300 3·000 2·950 6·250 less pot 300 opg. 900 gathg. 500 1·700 4·550	
	4·55		Flue ling. clay reb. circ. to BS 1181 type 2, 190 int. diam., jtd. & ptd. to 225 x 225 bk. flue in high alumina ct. mo.	Flue linings are measured in metres, with dimensioned descriptions (SMM F30.11.1.0.0). Cutting to form bends and easings, and cutting to walls around linings are deemed to be included (SMM F30.C7).
			Fender wall, etc. 1·350 2/400 800 2·150	No deduction of brickwork is made for voids ≤ 0·25 m² (SMM F10.M2b).
	2·15 0·30		Bk. wall, thickness: 215, comms. in English bond in c.m. (1:3).	Measurements of brick walls below hearth level are scaled from the plan.
			2·150 less 2/102·5 205 1·945	
	1·95 0·08		Bk. wall, thickness: 102·5, comms. in stret. bond in c.m. (1:3). (top of fender wall)	Walls supporting hearth and floor joists are measured in accordance with SMM F10.1.1.1.0.
	0·67 0·38		(support to back of hth.)	
2/	0·34 0·30		(support to fl. jsts. at sides of breast.)	

5.4

CHIMNEY BREAST, STACK AND FIREPLACE (Contd.)

	Dimensions	Description	Notes
	2.15 0.22 0.67 0.10	Dpc, width ≤225, hor., 2 cos. of slates, a.b.	Damp-proof courses are measured in m^2 in the appropriate classification as SMM F30.2.1.3.0.
2/	0.34 0.10		
2/	0.44 0.34 0.15	Ddt. Fillg. to make up levs., av. thickness ≤ 0.25m, obtained off site, selected gravel rejects.	Deduction of hardcore filling, compaction of filling, polythene membrane and concrete bed is required for the area occupied by the piers to the chimney breast, as although small quantities they adjoin the boundaries of measured areas (SMM General Rules 3.4).
2/	0.44 0.34	Ddt Compactg. fillg. & Ddt Damp-prf. memb., hor., polythene, a.b. & Ddt Trowellg. surf. of conc.	Hardcore filling is measured in accordance with SMM D20.10.1.3.1, the damp-proof membrane in accordance with SMM J40.1.1.0.0, and concrete bed as SMM E10.4.1.0.5.
2/	0.44 0.34 0.15	Ddt In situ conc. bed (1:2:4/20agg.), thickness ≤ 150. Hearth 338 102.5 235.5	
2/	0.90 0.40 0.38	Fillg. to make up levs., av. thickness > 0.25m, obtained off site, selected gravel rejects. (frt. hth.	Hardcore filling is measured in m^3 giving the particulars listed in SMM D20.10.2.3.1.
	0.67 0.24 0.38	(back hth.	
	0.90 0.40	Compactg. fillg. (frt. hth	Measured in m^2 as SMM D20.13.2.2.0.
	0.67 0.24	(back hth	

5.5

CHIMNEY BREAST, STACK AND FIREPLACE (Contd.)

Dimensions	Description	Waste / side calculations	Notes
		len. 900; 2/102.5 205; 1.105	Length scaled from plan.
1.11 0.50 0.15 0.67 0.34 0.15	In situ conc. bed (1:2:4/20 agg.), thickness ≤ 150.	(frt. hth. (back hth.	The concrete hearth is measured in m^3 as a bed, giving the appropriate thickness range as SMM E10.4.1–3. The compaction of the hardcore beneath has been measured previously, and hence it is not regarded as unblinded hardcore and does not therefore require specific mention.
2.11	Fwk. to edges of beds, plain vert., ht. ≤ 250.	1.105; add 2/500 1.000; 2.105	Formwork to edges to beds is measured in metres using the description and stating the depth classification as SMM E20.2.1.2.0.
1.11 0.50 0.67 0.34	Trowellg. surf. of conc.	fireplace 665; 2/338 676; 1.341 gathg. 2/500 1.000; 2/225 450; 1.450	Treating the surface of concrete is described and measured in m^2 (SMM E41.3.0.0.0).
1.34 0.90 1.45 0.60	Rendered coatg., width > 300, bk. back & sides to fireplace opg. & gatherg., ct. & sd. (1:3) in 1 ct. 15th., trowelled.		The location of the rendering should be given in the description as the working space is confined. This is permissible under SMM General Rules 1.1.
	Fireplace		Note: adjustment of flooring will be taken with floors.
1	Grate, 400 wide, & fret extensn. for all-night burng. basic price £48 (delvd. to site) & settg. in fireplace opg., inc. provsn. & fixg. of 400 solid one-piece fireclay fireback & all nec. fire ct. & conc. (1:10) backg. to fireplace opg., width: 570.		Provision and fixing of grate and concrete and brick backings are included in a single enumerated item. The only SMM item of any relevance appears to be SMM F30.16.1.1.0 (proprietary item). Note the use of basic price items to cover the cost of supply of fittings of this kind.
1	Fireplace surrd. 1.28 m wide x 1.00 m hi. w. 160 retn. & 400 fire opg. in pre-slabbed tile & faience, basic price £130 (delvd. to site) & setting & jtg. in fire ct. & Fireplace hth. 1.26 m x 250 x 85 ov'll. in pre-slabbed tile & faience basic price £60 (delvd. to site) & settg. & jtg. in fire ct.		The surround and hearth to the fireplace are best enumerated separately, giving the size, nature of material and condition in which the unit is to be supplied. Assembling and jointing or building up loose parts and setting should be given in the description.

5.6

RANDOM RUBBLE BOUNDARY WALL — EXAMPLE VI

(10 m long, 300 mm thick and 750 mm high, including rough stone coping)

Dimensions	Description	Notes
	750 less copg. 150 600	Rubble walls are measured in m^2 stating the thickness and plane (SMM F20.1.1.1.0), and giving full particulars of: 1. kind of stone 2. type of walling, including coursing 3. composition and mix of mortar 4. type of pointing (SMM F20. S1-6).
10.00 0.60	Stone wall, thickness: 300, random rubble, uncsd. in picked local sandst. from 'X' quarry, ro. dressed & bedded in l.m. (1:3), & ptd. b.s. in g.m. (1:1:6) w. flush jt.	Rough and fair raking or circular cutting are measured in metres stating the thickness (SMM F20.26-27.1.0.0). Labours in returns, ends and angles are deemed to be included (SMM F20. C1g), and the materials have already been measured.
10.00	Stone copg. 300 wide, av. 150 high, hor. ro. stones to match gen. wallg.	Levelling uncoursed rubble work for copings, damp-proof courses and the like is deemed to be included (SMM F20. C1j). Copings are measured in metres, giving a dimensioned description and the plane in which they are laid (SMM F20.16.1.3.0).

7 Measurement of Floors

SEQUENCE OF MEASUREMENT

It is essential to adopt a logical order in taking off this class of work to reduce the risk of omission of any items. The best procedure is probably to take each floor complete, starting with the highest floor in the building and working downwards.

The work on each floor can conveniently be sub-divided into: (1) construction and (2) finishings, and they are best taken in this sequence following the order of construction on the site. In the case of solid floors, the finishings may be measured in a finishings section which picks up all the items covered in the Surface Finishes Sections of the *Standard Method*.

HOLLOW FLOORS

Hollow floors consist of boarding nailed to timber joists. On a ground floor the joists will generally be supported on timber plates bedded on brick sleeper walls built off the concrete oversite. With upper floors the joists will be deeper and will usually be supported at their ends on walls or partitions, and some form of strutting will normally be incorporated at about 2 to 2.50 m centres, where the clear span of the joist exceeds 3 m.

The sequence of taking off should preferably follow the order of construction.

(1) Plates and bedding and possibly adjustment of brickwork (any supporting beams would also be taken at this stage).
(2) Floor joists.
(3) Strutting.
(4) Boarding.

Plates

Plates are measured in metres classified as plates with a dimensioned description and, where the length ≥ 6.00 m in a continuous length, the length is stated (SMM G20.8.0.1.1). Where brickwork has been measured over the plate and it measures 100 × 75 mm or more, then brickwork will be deducted in accordance with SMM F10.M3.

Floor Joists

The number of floor joists is obtained by taking the length of the room, from which 100 mm is deducted to allow for spaces between the end joists and walls, and dividing the adjusted length by the spacing of the joists, usually 350 to 450 mm. This gives the number of spaces to which one is added to give the number of joists.

Floor joists and beams are classified as floor members and measured in metres giving a dimensioned description in nominal sizes (SMM G20.6.0.1.0). Where joists are > 6.00 m in one continuous length, the length is stated. The building in of ends of timbers into brickwork is deemed to be included. It is necessary to state whether timbers are sawn or wrought (SMM G20.S1).

A worked example of the measurement of trimming floor joists around an opening is given in this chapter.

Joist Strutting

The most usual form of strutting of floor joists is herringbone strutting which, like solid or block strutting, is measured in metres over the joists, stating the depth of the joists and giving a dimensioned description (SMM G20.10.1–2.1.0).

Floor Boarding

Floor boarding is measured in m^2, giving a dimensioned description as SMM K20.2.1.1.0). Floorings in openings, although often $\leq 1.00\ m^2$ in area or $\leq$ 300 mm in width are not usually separated from the general flooring and enumerated or measured in metres respectively as SMM K20.2.1–3.1.0, since they are regarded as a natural extension of the general flooring. Bearers under flooring in openings are classified as floor members.

Doors are usually hung to open into rooms and so will normally be located on the room side of the wall. The floor finish in the door opening will accordingly generally be the same as that in the hall, passage or landing from which the room is entered.

Skirtings are usually taken at the same time as measuring the internal finishings, although they can be taken with the floors.

SOLID FLOORS

There are two main types of concrete floor construction

(1) Ground floors consisting of a concrete bed usually laid on a bed of hardcore.
(2) Upper floors consisting of suspended concrete slabs.

In each case the floor finishing can be measured at the same time as taking off the floor construction or be left to be measured with the other finishings. Concrete beds are often included with foundations.

Concrete Beds

In situ concrete beds are measured in m^3, stating the appropriate thickness range as SMM E10.4.1–3.0.5 and including in the description where poured on or against earth or unblinded hardcore. The prescribed thickness ranges are on ≤ 150 mm, 150–450 mm and > 450 mm. Treating the surface of *in situ* concrete is classified and given in m^2 as SMM E41.1–7.

Hardcore and similar beds are measured in m^3, classified as to whether the average thickness is ≤ or > 0.25 m, the nature of the filling material and its source and/or treatment (SMM D20.9–10.1–2.1–3.1–4).

Suspended Concrete Slabs

Suspended *in situ* concrete floor slabs are also measured in m^3 and classified in one of the three thickness stages given in SMM E10.5.1–3.0.0.

Chases formed in new brickwork to receive the edges of floor slabs are deemed to be included in the brickwork rates (SMM F10.Clc).

Bar reinforcement is billed in tonnes but entered on the dimensions paper in metres, keeping each nominal size, and straight, bent and curved bars, and links separate (SMM E30.1.1.1–4.1–4). Hooks and tying wire, and spacers and chairs where at the discretion of the contractor, are deemed to be included (SMM E30.C1). Horizontal bars, including those sloping ≤ 30°, with a length of 12.00 m or more, and vertical bars, including those sloping > 30°, with a length of 6.00 m or more, are each kept separate in 3.00 m stages (SMM E30.1.1.–3.1–4 and E30.D1–2).

Fabric reinforcement is measured in m^2, stating the mesh reference and weight per m^2 and minimum laps (SMM E30.4.1.0.0 and E30.S4).

Formwork to the soffits of floor slabs is measured in m^2 and the slab is classified as to thickness ≤ 200 mm and thereafter in 100 mm stages (SMM E20.8.1–2.1.0). The heights to soffits are classified as ≤ 1.50 m and thereafter in 1.50 m stages (SMM E20.8.1–2.1.1–2). Formwork to edges of slabs ≤ 1.00 m high is measured in metres, classified in the three height stages given in SMM E20.3.1.1–3.0, whereas that to edges > 1.00 m in height is measured in m^2.

Where beams are attached to slabs they are measured as part of the floor slab where their depth is ≤ three times their width (depth being measured below the slab) in accordance with SMM E10.D4a. Formwork to beams attached to slabs is measured in m^2, stating the number of beams in each item, the shape of the beam and the height to soffit in 1.50 m stages (SMM E20.13.1.1–2.1–2).

Floor Finishings

Where the finishings vary from one room to another it is usual to insert a dividing strip between the different floor finishings which is measured in metres with a dimensioned description in accordance with SMM M40.16.4.1.0.

Variations in the thickness of different floor finishings are generally overcome by varying the thickness of the screed, to maintain a uniform finished floor level throughout. Screeds are measured in m^2, giving the plane, thickness and number of coats (SMM M10.5.1.1.0).

WORKED EXAMPLES

Worked examples follow covering a variety of different forms of floor construction to ground and upper floors. In example X brief details are included of the measurement of reinforced *in situ* concrete and structural steelwork which it is hoped will be of value to the student.

The last worked example in this chapter covers the measurement of a timber partition, since this work is of similar character to that dealt with in timber floors.

HOLLOW GROUND FLOOR — EXAMPLE VII

Note: It is assumed that all earthwork has been measured previously, together with a cubic item of concrete bed (1:2½:5/20agg.), thickness ≤150mm and probably hardcore filling, average thickness ≤ 0.25m taken in m³. It is also assumed that the 75 mm concrete block partitions would be constructed off the concrete bed, and the latter will also need a trowelling item in m².

Timesing	Dimensions	Description	Side notes
3/	4.50 0.30	Bk. wall, thickness: 102.5, honeycombed, comms. in c.m.(1:3).	The brickwork is measured up to the underside of the plate, using the classification given in SMM F10.1.1.1.0.
3/	4.50 0.10	Dpc, width ≤ 225, hor., single layer of hessian-based bit. to BS 743 ref. A, bedded in c.m. (1:3).	Damp-proof course on sleeper walls, measured in m² and classified as ≤225 mm wide as SMM F30.2.1.3.0. It is assumed that the air bricks have been measured with the external walls, in the manner outlined in Chapter 6.
3/	4.50	Plates, 100 x 75, sn. swd. (wall plates	Plates are measured in metres giving a dimensioned description in accordance with SMM G20.8.0.1.0. It is necessary to state whether the timber is sawn or wrought (SMM G20.S1).
		Flr. Jsts. 4.500 less end spaces 2/50 100 400)4.400 11+1	The number of joists has been calculated to show the method of approach, although it could have been counted off the drawings in this case. 11 represents the number of spaces which 1 is added to give the number of the joists.
12/	3.60	Flr. members., 50 x 100, sn. swd. (jsts.	Joists are measured in metres as floor members, giving a dimensioned description in accordance with SMM G20.6.0.1.0. Some surveyors may still prefer to commence the description with the timber dimensions, ie '50 x 100 in sn. swd in floor members' and there is no universal practice. However, the author considers that the sequence shown in SMM should ideally be followed to secure uniformity of approach.

7.1

TIMBER FLOORS

Drawing No. 6

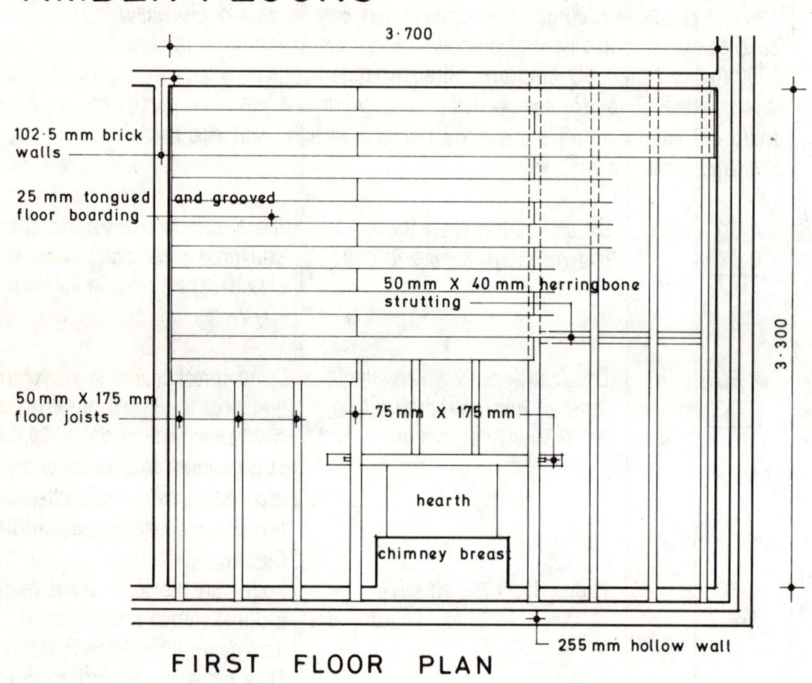

FIRST FLOOR PLAN

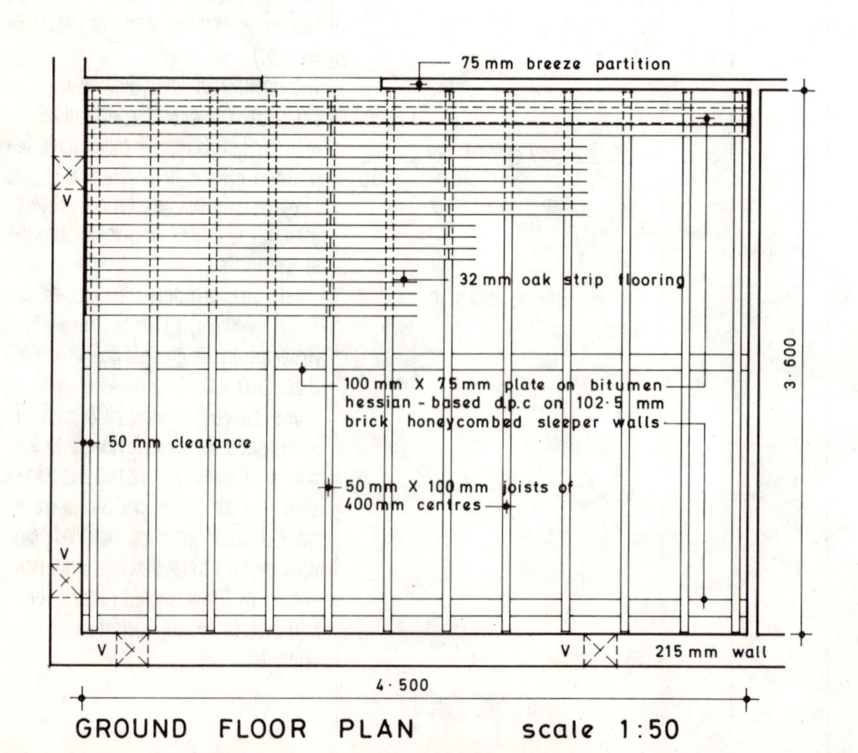

GROUND FLOOR PLAN scale 1:50

HOLLOW GROUND FLOOR (Contd.)

	4.50 3.60 0.75 0.08	Tbr. bd. flrg., width > 300, thickness: 32, wrot European oak t&g. in 75 widths, secret nailed to swd. jsts., & spld. headg. jts. (dr. opg.	Boarded flooring is measured in m² where > 300 mm wide, giving the particulars listed in SMM K20.2.1.1.0. If the thickness given is to be the finished thickness this must be stated (SMM K20. D2). Descriptions are to include the method of jointing or construction where not at the discretion of the contractor, and the nature of the background (SMM K20.S2-3). The boarding widths, being rather unusual and more costly, are included.
3/	0.08	Flr. membrs., 50 x 75 sn. swd. (brrs. in. dr. opg.	The flooring in door openings, although ≤ 1 m² in area or 300 mm in width, is included with that in the room as being a natural extension of it and not a disconnected or isolated area. The bearers in the opening are measured as floor members, as there is no separate provision in SMM7 for short lengths.

SOLID GROUND FLOOR — EXAMPLE VIII

(Alternative construction to the area covered in Example VII and illustrated in Drawing No. 6.)

Dimensions	Description	Notes
4.50 3.60	Compactg. bott. of excavn.	Compacting the surface of the ground to receive hardcore is a measured item in accordance with SMM D20.13.2.3.0.
4.50 3.60 0.10	Fillg. to make up levs., av. thickness ≤ 0.25 m, obtained off site, gravel rejects, blinded w. hoggin.	Hardcore filling is measured in m³ and giving the particulars listed in SMM D20.10.1.3.1.
4.50 3.60	Compactg. fillg.	Compacting filling item as SMM D20.13.2.2.0.
4.50 3.60 0.15 0.75 0.08 0.15	In situ conc. bed (1:2:4/20 agg.), thickness ≤ 150. (dr. opg.	Concrete beds are measured in m³, giving the thickness classification as SMM E10.4.1.0.0 and mix details as SMM E10.S1. The concrete surface will be lightly tamped to receive the screed and this is not a measured category in SMM E41.1–7, covering worked finishes.
4.50 3.60 0.75 0.08	Damp prfg., hor., polythene 1000 gauge, ld. on conc. (dr. opg.	It is assumed that the waterproof membrane under the screed will connect with the horizontal damp-proof course in the walls, otherwise a vertical strip would have to be measured. The damp-proof membrane is measured in accordance with SMM J40.1.1.0.0. and stating the nature of the base on which it is applied (SMM J40.S2). The floor finish of oak blocks on a floated bed has been taken with the floor slab, to show the method of measurement. Alternatively, the floor finishing could be taken with finishings generally to floors, walls and ceilings.
4.50 3.60 0.75 0.08	Screed to flr., lev., thickness: 25, one ct., fltd., ct. & sd. (1:3) on conc. bed (m/s). (dr. opg.	Screeds are measured in accordance with SMM M10.5.1, giving the composition and mix of materials, method of application and nature of base as SMM M10.S1–5. The screed has been taken as 25 mm thick to give a total thickness of floor finishing (blocks + bed) of 50 mm.

SOLID GROUND FLOOR (Contd.)

Timesing	Dimensions	Squaring	Description
	4·50 3·60		Wd. blk. flrg., lev., 300 x 75 x 25 European Oak, dipd. & jtd. in hot bit. on fltd. scrd. (m/s), dowelled, ld. herringbone w. 2 block plain margin a/rd., & sealg. & polishg. w. 2 cts. of wax pol.
	0·75 0·08		(dr. opg.
	0·75		Dividg. strip, 5 x 40 ebonite bedded in c.m. (1:3).

The last two items could be grouped against a single set of dimensions, and they have been separated here mainly to allow space for explanatory notes.

If other floor finishings of different thicknesses are involved, the screed thickness can be varied to give a uniform finished floor level throughout.

The wood block flooring is measured in m² in accordance with SMM M42.5.1.2.1. Full particulars of wood block flooring are given as required by SMM M42.S1-8. Plain block borders have been included in the description of the flooring, as they form part of the layout of joints. The nature of the finished surface, including sealing and polishing, is included in the description of the wood blocks.

It is normally feasible to omit the use of millimetres when entering the dimensions of components in descriptions and this is increasingly becoming standard practice.

Provided in door openings to separate different floor finishings, measured in accordance with SMM M40.16.4.1.0.

BOARDED UPPER FLOOR

EXAMPLE IX

		len. 3·300 add beargs. 2/100 200 3·500	Bearings added to give total length of floor joists; taken as 100 mm each to give a slight clearance and less risk of moisture absorption from the cavity.
10/	3·50	Flr. members. 50 x 175 sn. swd. (jsts.	The number of joists can be counted direct from the drawing and need not be calculated. The classification follows SMM G20.6.0.1.0. Note the order of dimensions in the descriptions of width and depth.
2/ 2/	0·90 3·50	Ddt ditto. (trimd. jsts. (trimg. jsts.	Adjustment for shorter lengths of trimmed joists, allowing for jointing to trimmer, and deduction for trimming joists.
		add 1·100 trimg. jsts. 2/75 150 tusk tenons 2/150 300 1·550	Note build-up of length of trimmer, including an allowance of 150 mm for each tusk tenon, although they are unlikely to be used in practice, galvanised steel hangers being a more likely approach.
2/	3·50 1·55	Add ditto. 75 x 175 (trimg. jsts. (trimmer	Substitution of thicker trimming joists and addition of trimmer, all included under the general classification of floor members. No mention has been made of the jointing as it is at the discretion of the contractor (SMM G20.S9).
	3·70	Jst. struttg. 50 x 40 sn. swd., herringbone, 175 dp. jsts.	Measured in metres over the joists and giving the size of strutting and depth of joists as SMM G20.10.1.1.0 and G20.M1. Where the ends of joists are supported by hangers, these are enumerated giving a dimensioned description or diagram (SMM G20.21.1–2).
2/	0·50	Flr. membrs. 50 x 75 sn. swd. (cradlg. pieces)	Bearers at the ends of hearth are measured in accordance with SMM G20.6.0.1.0.

BOARDED UPPER FLOOR (Contd.)

Dimensions	Description	Notes
3.70 3.30	Tbr. bd. flrg., width > 300, thickness: 25, wrot. swd. t.& g. fxd. w. flr. brads to swd. jsts. (m/s), & spld. headg. jts., as specfn. clause —.	Measured in m^2 with the particulars listed in SMM K20.2.1.1.0 and K20.S1. In practice boarding will also be needed to a door opening to this room.
0.90 0.34 0.85 0.53	Ddt ditto. (chy. breast (hth. 850 2/530 1.060 1.910	Deductions of flooring for chimney breast and hearth, including margins, which are measured separately to the floorboarding. Note that the void allowance in SMM K20.M1 ($0.50\ m^2$) does not apply to voids adjacent to boundaries of measured areas (SMM General Rules 3.4). No cutting to flooring around openings is measurable as labours are deemed to be included (SMM K20.C1a).
1.91	Tbr. bd. flrg., width ≤ 300, 50 x 25 wrot European Oak margin.	Margins around hearths are measured separately in metres as width ≤ 300 mm. It must be measured separately from the general boarding as it is of different timber.

9.2

CONCRETE UPPER FLOOR — EXAMPLE X

Alternative construction to the area covered in Example IX illustrated in Drawing No. 6.

Dimensions	Description	Notes
	len. 3·700, width 3·300; add bearg s. 2/100: 200, 200; = 3·900, 3·500	
3·90 3·50 0·15	In situ conc. slab (1:2:4/20 agg.), thickness ≤ 150, reinfd.	In situ concrete slabs are measured in m^3 stating the thickness range and reinforced classifications as SMM E10. 5.1.0.1 and the mix as SMM E10.S1.
0·90 0·34 0·15	Ddt. ditto. (chy. breast	
	3·900, 3·500; less cover 2/40: 80, 80; = 3·820, 3·420	It is usual to allow about 40 mm cover to all reinforcement to prevent the possible onset of corrosion.
3·82 3·42	Stl. fabric reinft. to BS4483 ref. A 193, weighg. 3·02 kg/m^2; w. 100 min. laps.	Fabric reinforcement is measured in m^2 giving the particulars listed in SMM E30.4.1.0.0 and E30.S4. Tying wire, cutting, bending, and spacers and chairs which are at the contractor's discretion are deemed to be included (SMM E30.C2).
	900; add cover 2/40: 80; = 980	Forming chases in brickwork to receive the concrete slab are deemed to be included in the brickwork rates (SMM F10. C1c).
0·98 0·34	Ddt. ditto. (chy. breast	
3·70 3·30	Fwk, soff. of slab, slab thickness ≤ 200, hor., ht. to soff. 1·50–3·00 m.	Formwork classified as SMM E20.8.1.1.2, according to the concrete slab thickness and height to soffit. Where the thickness of the slab > 200 mm, the formwork shall be given separately in 100 mm stages of slab thickness, while the height to soffit of slab is given in 1·50 m stages.
0·90 0·34	Ddt. ditto. (chy. breast	
3·90 3·50	Trowellg. surf. of conc.	Trowelling the surface of concrete is so described and measured in m^2 (SMM E41.3.0.0.0).
0·90 0·34	Ddt. ditto. (chy. breast	

CONCRETE UPPER FLOOR (Contd.)

Method of Measurement of In situ Reinforced Concrete Beam (225 x 225)

(If required to give intermediate support)

Timesing	Dimensions	Description	Side notes
		len. 3·700; add beargs. 2/100 200; 3·900	In situ concrete attached beams to suspended slabs are included with the slabs, except where they are deep beams, with a depth/width ratio exceeding 3:1, measured below the slab (SMM E10. D4a). Although this results in a thicker slab over a small area, it is considered permissible to include the beam concrete in with the slab concrete, without any change of thickness classification. Furthermore, SMM E10. M2 appears to support this approach as it states that the thickness of slabs stated in descriptions excludes projections and recesses. Indeed it would be rather unrealistic to measure a small section of slab of the combined thickness in the range 150 - 450 mm thick.
	3·90 0·23 0·23	In situ conc. slab (1:2:4/20 agg.), thickness ≤ 150, reinfd. (beam under slab len. 3·900; less cover 2/40 80; 3·820; add hkd. ends 2/300 600; 4·420	
3/	4·42	M.s. bar reinft. 25φ nom. size, strt., to BS4449. gth. of beam: sides 2/225 450; soff. 225; 675	It is assumed that the beam is reinforced with 3nr. 25 mm diameter mild steel bars. The length of the bars is obtained by taking the full length of the beam, deducting 40 mm cover at each end and adding twelve times the diameter for each hooked end. The reinforcement will subsequently be reduced to tonnes. It is measured in accordance with SMM E30.1.1.2.0, with each diameter given separately under the classification of straight bars.
	3·70 0·68	Fwk. to beams, attached to slab, reg. sq. shape, ht. to soff. 1·50 - 3·00 m (In 1 nr.).	Formwork to attached beams is measured in m^2 stating the number of members, shape and height to soffit (SMM E20.13.1.1.2).
	3·70 0·23	Ddt Fwk., soff. of slab., a.b.	Deduction of formwork to soffit of slab for area occupied by beam.

(Note: It is assumed that the floor finishing is taken in the Finishings section. See Chapter 9 for further examples of the measurement of floor finishings)

10.2

CONCRETE UPPER FLOOR (Contd.)

Method of measurement of steel joist (203 x 133), encased with concrete, as an alternative to the reinforced concrete beam.

Dimensions	Description	Notes
1	Stl. isoltd. structl. membr., plain beam, 203 x 133 x 30 kg/m, 3.90 m lg. to BS 4 Pt. 1, table 5, supptg. conc. slab.	Isolated structural steel members are billed in tonnes to classifications in SMM G12.5.1.2.0, including use and details of construction. Any fittings would only be measured separately, if they were of a different type and grade of material (SMM G12.M2). Building in ends of steel members are not measured but if padstones were required, they would form a separate enumerated item.
	width 133, depth 203; add cover 75, 40; = 208, 243	Build up of dimensions of in situ concrete in casing to steel beam below slab.
3.90 0.21 0.24	In situ conc. slab (1:2:4/20 agg.), thickness ≤ 150, reinfd.	The concrete in the suspended slab is deemed to include concrete casings to steel beams (SMM E10.D4), but giving the suspended slab thickness range classification as in the case of the reinforced concrete beam.
	gth. of beam: sides 2/243 = 486; soff. 208; = 694	
3.70 0.69	Fwk. to beam casg. attached to in situ slab, reg. rect. shape, ht. to soff. 1.50 – 3.00 m (In 1 nr).	Formwork to attached beams is measured in accordance with SMM E20.14.1.1.2. This will be followed by adjustment of the formwork to soffit of slab item as in previous example.

STUD PARTITION

Drawing No. 7

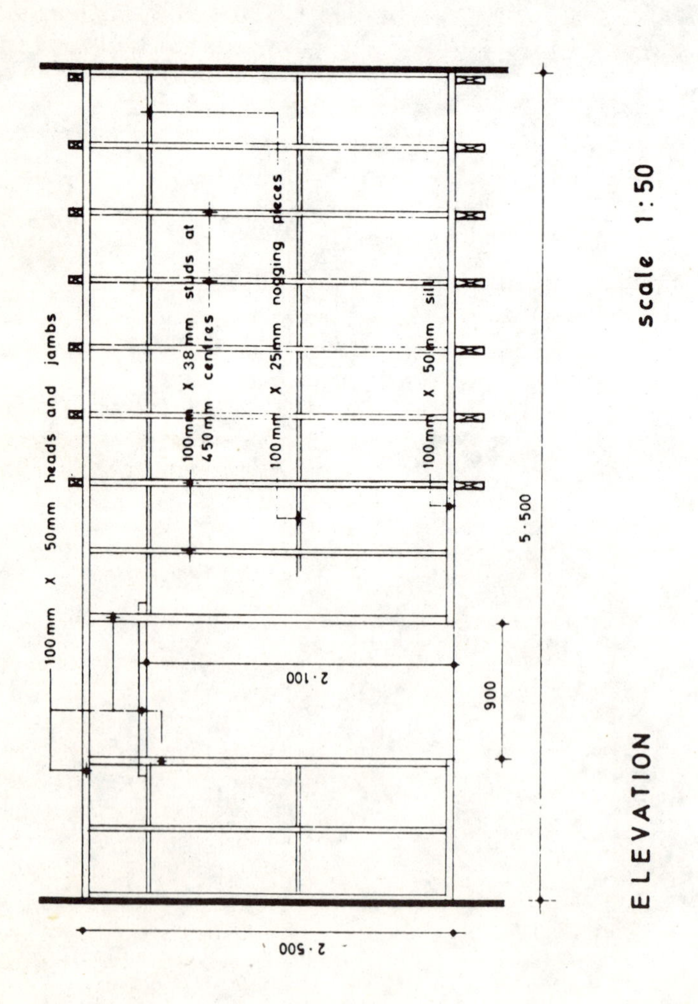

STUD PARTITION

EXAMPLE XI

		ht. 2·500 less hd. & sill 2/50 100 2·400 add stub tenons 2/25 50 2·450	Build-up of length of studs, assuming that the studs are stub-tenoned into the head and sill.
10/	2·45	Ptn. membrs. 100 x 38 sn. swd. (studs	Studs are measured in metres as partition members, giving a dimensioned description (SMM G20. 7.0.1.0).
2/	2·50	Ditto. 100 x 50. (dr. jambs	It is necessary to distinguish between sawn and wrought timbers in all cases (SMM G20.S1).
		900 add jambs 2/50 100 t.tenons 2/75 150 1·150	Note build-up of length of door head in waste.
	1·15	Ditto. 100 x 50. (dr. head	
		5·500 less dr. opg. 900 4·600	
	4·60	Ditto. 100 x 50 (sill	The length of the sill is obtained by deducting the width of the door opening from the length of the partition.
	5·50	(head	
		less 5·500 dr. opg. 900 dr. jbs. 2/50 100 studs 10/38 380 1·380 4·120	
2/	4·12	Ditto. 100 x 25 (noggings	Partition members include struts and noggings (SMM G20.D3). Hence noggings are measured their actual lengths as a linear item like other partition members. There is no provision for them to be measured in a similar manner to joist strutting as SMM G20.10.1–2 and G20.M1. Plugging of end studs is not described with the items as it is assumed that the method of fixing is at the discretion of the contractor (SMM G20.S2). It is assumed that the linings, skirtings and decorations are measured with Internal Finishes.

11.1

8 Measurement of Roofs

Both flat and pitched roofs can conveniently be subdivided into two main sections for purposes of measurement, that is, construction and coverings. The order of measurement of these two sections varies in practice, but on balance it is probably better to take the construction first since this follows the order of erection on the site.

The general rules for the measurement of the lengths of rafters, hips and valleys were described in chapter 3. The work to the eaves of pitched roofs and rainwater goods is normally measured with the roofs.

With flat roofs the taker-off must ensure that all rolls, drips, gutters and associated work are drawn on the roof plan before he starts taking-off, otherwise his dimensions may be woefully deficient.

PITCHED ROOFS

Roof Timbers

The order of items should follow a logical sequence such as plates, rafters, ceiling joists, collars, purlins, struts, ridge boards and hip and valley rafters.

When determining the length of the plate, it is customary to allow for laps at corners equal to the width of the plate, but not for laps in running lengths. Where joints are not described they are at the discretion of the contractor (SMM G20.S9). Deduction of brickwork for the volume occupied by the plate will be needed where the brick walling has been previously measured and the plate has a minimum size of 100 × 75 mm (SMM F10.M3).

The roof timbers are measured in metres as roof members, stating the pitch and giving a dimensioned description, and these include struts, purlins, rafters, hip and valley rafters, ridge boards, ceiling joists, binders and bracing (SMM G20.9.2.1.0 and G20.D6). It is necessary to add a rafter for each hipped end, but apart from this the number of rafters will be the same as if the roof were gabled. Where the length of a member is > 6.00 m in one continuous length, the length is stated. Plates are classified as such in accordance with SMM G20.8.0.1.0.

The adjustment of roof timbers around chimney stacks is usually taken after the measurement of the main roof timbers and coverings, and a worked example is provided in example XIII.

Roof trusses, trussed rafters and trussed beams are enumerated and provided with a dimensioned description as SMM G20.1–3.1.0.0. The work is deemed to include webs, gussets and the like (SMM G20.C2).

Coverings

The area of slating or tiling is measured first in m^2 stating the pitch, and is deemed to include underlay and battens (SMM H60.1.1.0.0 and H60.C1). Note the various particulars that are required in the billed description of slating or tiling, including the kind, quality and size of slates or tiles, method of fixing, minimum laps and spacing of battens (SMM H60.S1–4).

These items are followed by linear roofing boundary items, such as work at abutments, eaves, verges, ridges, hips, vertical angles and valleys, and describing the method of forming them (SMM H60.3–9). Boundary work to roofs is deemed to include undercloaks, cutting, bedding, pointing, ends, angles and intersections (SMM H60.C2).

These measurements will be followed by the adjustment of roof coverings for chimney stacks or other openings through the roof. Where the opening is ≤ 1.00 m^2, no deduction will be made from the area of roof covering (SMM H60.M1), and no boundary work will be measured to the void (SMM H60.M2).

Eaves and Rainwater Goods

Eaves or verge soffit boards and fascia boards (including barge boards), and gutter boards (including sides) are each measured separately in metres where the width is ≤ 300 mm, with a dimensioned overall cross-section description. Where the width is > 300 mm, they are measured in m^2 with a dimensioned description (SMM G20.14–16.1–3.1–2.0). Adjustments will have to be made to the length measured on the outside face of the external walls for corners to arrive at the correct girth of fascia board. The student must also be sure to deduct the width of gables, where no fascia is required.

The supporting bearers to eaves and soffit boarding are measured by length as framed supports where ≤ 300 mm wide, giving a dimensioned overall cross-section description and spacing of the members (SMM G20.12.2.1.4). Framed supports are where the members are jointed together other than butt jointed (SMM G20.D8). Remember to take the painting of the exposed timber surfaces as general surfaces (SMM M60.1.0.1–2.0).

Both downpipes and eaves gutters are measured in metres over all fittings and branches, with the fittings, made bends and special joints and connections enumerated as extra over the pipe or gutter in which they occur (SMM R10.1.1.1.1, R10.10.1.1.1, R10.2.1–4.1–5.1, R10.11.1–2.1.1 and R10.M1 and M6). Pipe and gutter types, nominal sizes, method of jointing, and type, spacing and method of fixing supports shall be included in the descriptions of pipes and gutters. The pipe and gutter items will be followed by their painting (SMM M60.8.2.2.0 and M60.9.0.2.0). Painting work on surfaces > 300 mm girth is given in m^2, with that to narrower isolated surfaces in metres.

Note: Profiled sheet roofing is covered in SMM work sections H30 (fibre cement), H31 (metal), H32 (plastics) and H33 (bitumen and fibre).

FLAT ROOF COVERINGS

Asphalt

The rules for the measurement of asphalt coverings to flat roofs are detailed in SMM work section J21. The main areas of asphalt roofing are measured in m^2, stating the pitch. The particulars listed in SMM J21.S1–4 must be included and these cover the kind, quality and size of materials, thickness and number of coats, nature of base on which applied and surface treatments. Furthermore, the asphalt item is deemed to include cutting, notching, bending and extra material for lapping the underlay and reinforcement, and working the asphalt into recesses (SMM J21.C1b and c). Linear items will follow the covering, such as skirtings, fascias and aprons, linings to gutters, channels and valleys, coverings to kerbs, internal angle fillets and appropriate labours (SMM J21.5–17), followed by any enumerated items such as collars around pipes, linings to cesspools and sumps, and roof ventilators (SMM J21.18–21 and 23).

Built up Felt

Built up felt roof coverings are measured in m^2, and stating the pitch as SMM J41.2.1.0.0. Full particulars of the felt are to be given including the kind and quality of felt, nature of base on which applied and method of jointing (SMM J41.S1–3). The measurement of the main areas of roof covering will be followed by such linear items as are appropriate, giving the girth in 200 mm stages (SMM J41.3–14.2.0.0), finishing with any enumerated items (SMM J41.15–18.1.0.0).

Sheet Metal

The measurement of sheet metal roofing in lead, aluminium, copper and zinc is dealt with in SMM work sections H71–74. The main areas of sheet metal are taken first, to be billed in m^2, stating the pitch, and allowing for the additional material at drips, welts, rolls, seams, laps and upstands/downstands, in accordance with the allowances given in SMM H71–74.M2. Full particulars of the sheet metal must be given as required by SMM H71–74.S1–6, including the underlay, thickness, weight and temper grade, method of fixing, details of laps, drips, welts, rolls, upstands and downstands and the like, and type of support materials and special finishes.

Flashings, aprons and cappings and the like are measured in metres with a dimensioned description, and stating whether horizontal, sloping, vertical, stepped or preformed (SMM H71–74.10–18.1.0.1–5). Gutters and welted, beaded and shaped edges are also measured as linear items. Saddles, soakers and slates, hatch covers and ventilators are each enumerated with a dimensioned description (SMM H71–74.25–28.1.0.0).

WORKED EXAMPLES

Worked examples follow covering the measurement of a tiled pitched roof, adjustment of roofing for a chimney stack, and asphalt and lead covered flat roofs.

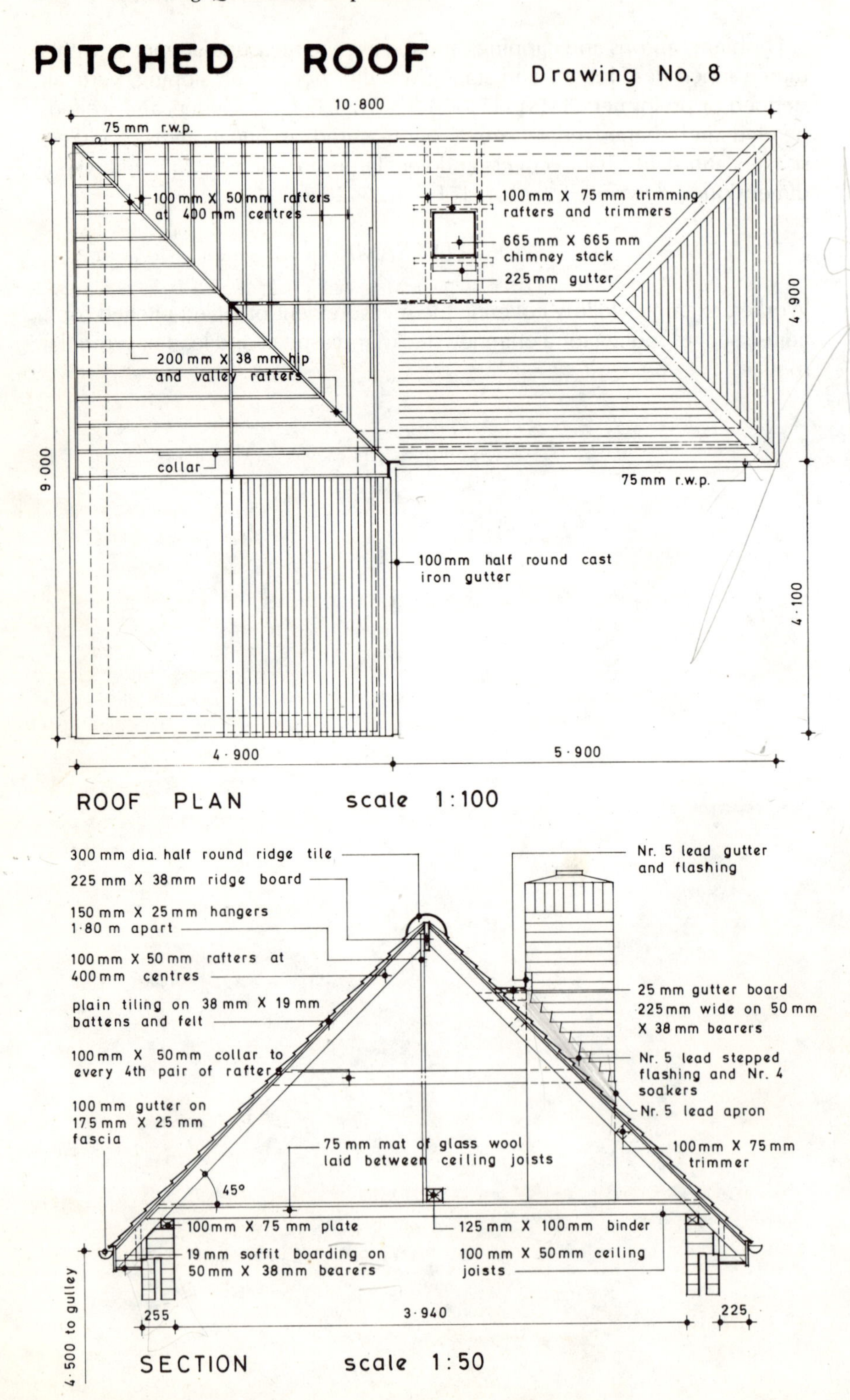
PITCHED ROOF
Drawing No. 8
10·800
75 mm r.w.p.
100 mm X 50 mm rafters at 400 mm centres
100 mm X 75 mm trimming rafters and trimmers
665 mm X 665 mm chimney stack
225mm gutter
4·900
200 mm X 38 mm hip and valley rafters
9·000
collar
75 mm r.w.p.
100mm half round cast iron gutter
4·100
4·900
5·900
ROOF PLAN scale 1:100
300 mm dia. half round ridge tile
225 mm X 38mm ridge board
150 mm X 25 mm hangers 1·80 m apart
100 mm X 50 mm rafters at 400 mm centres
plain tiling on 38 mm X 19 mm battens and felt
100 mm X 50 mm collar to every 4th pair of rafters
100 mm gutter on 175 mm X 25 mm fascia
Nr. 5 lead gutter and flashing
25 mm gutter board 225mm wide on 50 mm X 38 mm bearers
Nr. 5 lead stepped flashing and Nr. 4 soakers
Nr. 5 lead apron
100mm X 75 mm trimmer
75 mm mat of glass wool laid between ceiling joists
45°
100mm X 75 mm plate
125 mm X 100mm binder
19 mm soffit boarding on 50 mm X 38 mm bearers
100 mm X 50mm ceiling joists
4·500 to gulley
255
3·940
225
SECTION scale 1:50

PITCHED ROOF

EXAMPLE XII

The order of taking off has been taken as: 1. Roof timbers.
2. Coverings.
3. Eaves and rainwater goods.

With the roof timbers, adopt a logical sequence of taking off, such as plates, rafters, ceiling joists, collars, purlins, ceiling beams, struts, ridge board and hip and valley rafters.

Note detailed build-up of length of plate, making allowance for the gable and laps at corners, measuring to the extremities of the timber. There is however no allowance made for laps in running lengths.

In this case the external angle does not cancel out the internal angle, as shown below. The internal angle is picked up by measurements on the internal face of the wall, but lapped lengths at the hatched external angle still remain to be taken.

Roof tbrs.

	Plate
	10·800
less o'hg. at eaves 2/225	450
	10·350
	9·000
less eaves & verge	300
	8·700
	10·350
	2/ 19·050
	38·100
less crnrs. 4/2/255	2·040
len. on int. face	36·060
less gable	3·940
	32·120
add laps at crnrs. 3/2/100	600
	32·720

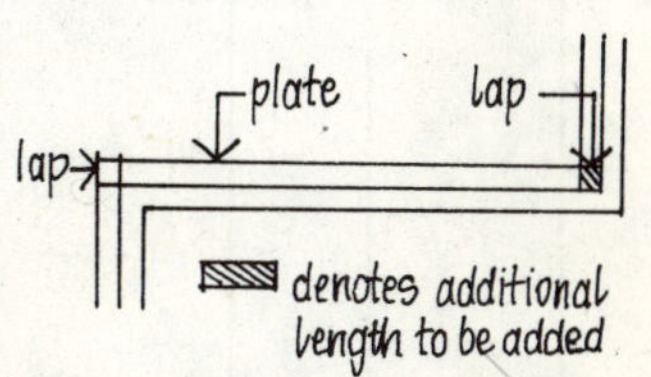

(2 lapped lengths to be taken at each corner)

Dimensions	Description	Notes
32·72	Plate, 100 x 75 sn. swd. (wall plate	Classified as SMM G20.8.0.1.0 with a dimensioned description.
	32·120	
	add crnrs. 2/215 — 430	
	32·550	
32·55 0·08	Ddt. Bk. wall thickness: 215, comms. in g.m. (1:1:6).	Brickwork to be deducted for volume occupied by plate as SMM F10.M3.
	width of 1B wall — 215	Thus it is necessary to deduct the one-brick wall, previously measured, and to add back the half-brick wall on the outer face.
	less ½ width of ½B wall — 51	
	164	
	32·120	
	add crnrs. 2/2/164 — 656	
	32·776	

12.1

PITCHED ROOF (Contd.)

	32.78 0.08		Add Bk. wall thickness : 102.5, comms. in g.m. (1:1:6).	The number of pairs of rafters in the main length of the roof is obtained from the number of spaces, with one added to convert spaces to pairs of rafters.
			Rafters 10.800 less o'hang 2/225 450 walls 2/255 510 end spaces 2/50 100 — 1.060 400) 9.740 25+1	
			less end space 50 — 4.100 gable wall 215 verge 75 — 340 400) 3.760 10+1 26 11 37	Then proceed to calculate the number of rafters in the bottom leg of the roof. Two are added for the extra rafters at the hips at each end of the roof. All roof timbers except plates are classified as roof members, pitched, giving a dimensioned description (SMM G20.9.2.1.0).
37/2/	3.54		Rf. membrs., pitched., 50 x 100 sn. swd.	The wording of the description follows the same sequence as in SMM7, but not all surveyors will necessarily adopt this approach.
2/	3.54		(ex. at hipd. ends	Lengths have been scaled off the drawing, but they could have been calculated as follows: Length of rafter = 1/2 overall span X Sec 45° + tapered end = 2.450 X 1.414 + 75 = 3.464 + 75 = 3.539
			Clg. jsts.	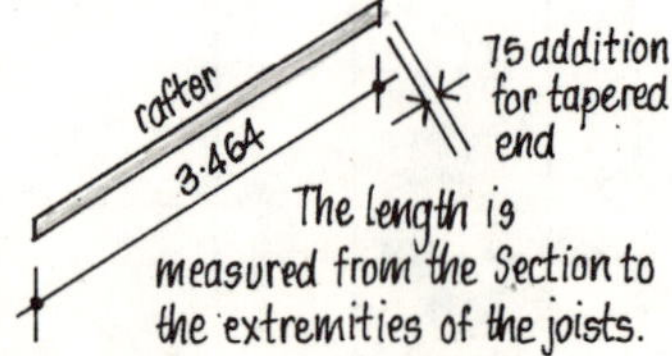
37/	4.35		Rf. membrs., pitchd., 50 x 100 sn. swd. (clg. jsts.	Classification as SMM G20.9.2.1.0.
	4.35		Ddt Ditto. 50 x 100 sn. swd. & Add Ditto. 50 x 130 sn. swd. (trimmer	A large member is needed to support the ends of ceiling joists across the intersection of the two roofs. All labours on members are deemed to be included (SMM G20.C1).
			Collars 4) 37 9	This calculation is rather approximate as the presence of the hipped corner prevents the use of a collar for some distance from the corner.
9/	2.35		Rf. membrs., pitched., 50 x 100 sn. swd. (collars	One collar is provided to each fourth pair of rafters.

PITCHED ROOF (Contd.)

Binders

binder

Timesing	Dimensions	Squaring	Description	Notes
			wall 255 o'hg. 225 480 wall 215 verge 75 290 10·800 less 2/480 960 → 9·840 4·100 480 4·580 less 290 → 4·290 14·130	The calculations for determining the lengths of binders is shown in waste. It is also assumed that no lengths exceeding 6·00 m will be required.
	14·13		Rf. membrs., pitchd., 100 x 125 sn. swd. (binders	No distinction is made between roof members of different types in bill descriptions. Trimming of roof timbers around the chimney stack opening will be taken later in adjustments for chimney stack (Example XIII).
			Ridge bd. 10·800 less 2/½/4·900 4·900 5·900	Calculation of length of ridge board is shown in waste.
	5·90		Rf. membrs., pitched., 38 x 225 sn. swd. (ridge bd.	Cutting and fitting ends of rafters against ridge board are deemed to be included, as they will be covered by SMM G20. C1. Ridge boards have been classified in the same way as other pitched roof members.
			9·000 less 290 8·710 add bearg. 100 8·810 less ½/4·900 2·450 6·360	
	6·36		Ditto. in one continuous len.: 6·36 m. (do	Members > 6·00m in one continuous length are kept separate and the length stated.
			Hangers 6·360 5·900 less 100 6·260 1·800)12·160 7+1	Hangers are spaced 1·80 m apart. The length is scaled from the Section. The adjusted ridge board length is used to calculate the number of hangers.
8/	2·00		Rf. membrs., pitched., 150 x 25 sn. swd. (hangers	The length of hip rafter is found by plotting the height of the roof on plan at the hipped end and drawing the hip to the slope of the roof as illustrated in Chapter 3 and adding an allowance for a tapered end. (alternatively the length can be calculated).
			4·250 add tapd. end 75 4·325	
4/	4·33		Rf. membrs., pitchd., 38 x 200 sn. swd. (hip & valley rafters	Roof trusses are enumerated with a dimensioned description (SMM G20.1.1).

PITCHED ROOF (Contd.)

	Dimensions	Description	Notes
		Insulation len. of side rf. 4·100; add intersectn. 255 225 480 = 4·580; less gable 215 75 290 = 4·290	Build-up of areas of roof insulation in waste.
		len. of main rf. 10·800 less 2/480 960 = 9·840; add 2/0'lap on wall 2/120 240 = 10·080. width 4·900 less 960 = 3·940; add 240 = 4·180	
	10·08 4·18 4·29 4·18	Insulatn. quilt; hor., glass wool to BS 1785, 75th., (bit. bonded), w. 100 laps, between members @ 400 ccs. (mesd. o/a).	Insulation quilts are measured the area covered in m² (SMM P10.M1) and giving the particulars listed in SMM P10.2.3.1.0 and P10.S1-3. No deduction is made for the thickness of ceiling joists as the strips of insulation quilt usually extend a little way up the sides of the joists, to ensure complete coverage. The addition of (measured overall) should clarify the position.
		Roof Covergs. 10·800 add proj. of tilg. over gutter 2/40 80 = 10·880	
2/ 2/	10·88 3·50 4·10 3·50	Rf. covergs. to 45° pitch, 267 x 165 clayware mach. made plain rfg. tiles to BS402 basic price £180 per thsd. (delvd. to site) double lapped w. 65 min. lap & nailed w. 2 nr. 40 lg. compo nails in every 4 th. cos. & at eaves on & inc. 38 x 19 sn. swd. battens, treated w. preservative to BS 4072, @ 100 ccs, fxd. w. galvd. nails, & underfeltg. of bit. felt (type 1a) to BS747, w. 150 min. laps & fxd. w. galv. clout nails.	Roof tiling is measured in m² stating the pitch (SMM H60.1.1.0.0) and giving the particulars listed in SMM H60.S1-4, including the kind, quality and size of materials, method of fixing, minimum laps and spacing of battens. The same rules apply to slates. The length on slope of the tiling has been scaled from the section, and is the same length as the rafters. Coverings are deemed to include underlay and battens (SMM H60.C1). The areas of roof coverings at hips and valleys balance one another. Note the different form of construction for single lap tiling illustrated in 'Building Technology' by the same author.
		Eaves 10·880, 9·040 = 19·920, 2/ = 39·840, less gable 4·900 = 34·940	
	34·94	Eaves to tile rf. coverg., inc. 75 x 39 (extr.) sn. swd. tiltg. fillet.	Eaves are measured in metres (SMM H60.4), and the eaves tilting fillet has been included in the item.
		Ridge len. 5·900, 6·260, gable wall 290 = 12·450	Adopt a logical order for linear roof covering items to boundaries such as eaves, ridge, hips, valleys and verges.

12.4

PITCHED ROOF (Contd.)

Timesing	Dimensions	Squaring	Description	Notes
	12.45		Ridge tiles, 300dia. h.r. clayware, (basic price £190/100) to match gen. tilg. & b.&p. in c.m. (1:2).	Ridge tiles are measured in metres as SMM H60.6, and giving the kind and quality of tiles and type of fixing as SMM H60.S1-2. Boundary work is deemed to include undercloaks, cutting, bedding, pointing, ends, angles and intersections (SMM H60.C2).
3/	4.33		Hip tiles 300 dia. h.r. clayware, (basic price £190/100) to match gen. tilg., & b.&p. in c.m. (1:2).	Measured similar to ridge tiles, but keep as a separate item (SMM H60.7). The basic price per 100 tiles assists the estimator in pricing the item.
3/	1		Galvd. hip iron scrd. to ft. of swd. rafter.	Enumerated item as SMM H60.10.4.1.0 taken when hip coverings are unnailed to prevent the lower tiles slipping.
	4.33		Valley tiles, p.m. (basic price £200/100) nailed to battens.	Linear item as SMM H60.9.
2/	3.50		Verges, tile-&-a-half width tiles & ex. underclk. cos, inc. b&p. & fin. w. fillet in c.m. (1:2).	Verges are measured separately as SMM H60.5, and including particulars of method of forming (SMM H60.S5).
			Eaves & R.W. Gds.	Fascia boards are measured in metres, giving a dimensioned overall c.s. description (SMM G20.15.3.2.0).
	34.94		Fascia bd. 25 x 175 wrot swd.	Fascia has the same length as eaves in this case, but watch out for tile corbels at gables and the like.
			&	
			Paintg. gen. isoltd. surfs., ext., wd., girth ≤ 300, k.p.s. & (3).	Painting to fascia classified as to general isolated surfaces, wood, and measured in metres as ≤ 300 mm girth (SMM M60.1.0.2.0). The nature of the base is to be stated (SMM M60.S2) and painting is deemed to be internal, unless otherwise described (SMM M60.D1).
2/	1		Gusset end to eaves, irreg. shape, 225 x 250 (av.) x 25, wrot. swd.	Gusset ends are enumerated with a dimensioned description as SMM G20.18.0.1.0 and stating whether timber is sawn or wrought (SMM G20.S1).

12.5

PITCHED ROOF (contd.)

Timesing	Dimension	Squaring	Description	Notes
2/	0.23		Paintg. gen. isoltd. surfs., ext., wd., gth. ≤ 300, k.p.s. & (3). (boxed ends to eaves)	The painting to the boxed ends is added to that of general surfaces (fascia to which it is attached). Some surveyors may prefer to start the painting description with k.p.s. & (3) and various approaches can be adopted. It seems logical to extract the relevant particulars from the first four columns of SMM7 and to finish the description with the supplementary information contained in the last column, and the painting system could also constitute a specification reference.
			10.800 / 9.000 / 2/19.800 / 39.600 / less gable 4.900 / 34.700; add int. L's 2/200: 34.700 + 400 = 35.100; eaves proj. 225, less fascia 25 = 200	
	35.10		Eaves soffit bd. 200 x 19 wrot swd. t.&g. fxd. to swd. suppts. (m/s). & Framed suppts., width ≤ 300, 50 x 38 sn. swd @ 400 ccs. to irreg. shaped area, 200 x 350 o/a.	Measured in metres to the classification given in SMM G20.16.3.2.0. The bearers supporting the eaves are measured as framed supports in metres, giving the size and spacing as SMM G20.12.2.1.4. Framed supports are where the members are jointed together other than butt jointed (SMM G20.D8).
			len. 34.700, less angles 2/225 450 = 34.250	Length of eaves adjusted as painting is measured on centre line of eaves boarding. This is best kept separate from the fascia, even although the combined girth exceeds 300 mm, as they are likely to be finished in different colours.
	34.25		Paintg. gen. isoltd. surfs., ext. wd., gth. ≤ 300, k.p.s. & (3).	
			<u>Rainwater Installation</u> len. of gutter 34.700, add angles 2/100 200 = 34.900	
	34.90		Gutter, st., nom. size 100, h.r.c.i. to BS460, bolted tog. & jtd. in r.l. putty, on & inc. w.i. stand. bkts. at 1 m ccs. scrd. to swd. & Paintg. eaves gutter, isoltd. surfs, ext. met., gth. ≤ 300, prep. & (3).	The gutter is measured on its centre line over all gutter fittings (SMM R10.M6). The description of the gutter is to include the type, nominal size, method of jointing, and type, spacing and method of fixing supports and background (SMM R10.10.1.1.1). Alternatively the rainwater goods might be of unplasticised PVC to BS 4576, clipped together and supported by vinyl brackets. These are usually left unpainted. Follow the gutter with the painting item, as SMM M60.8.2.2.0. Painting work to gutters is deemed to include work to gutter brackets (SMM M60.C7).

12.6

PITCHED ROOF (Contd.)

2/	1	E.o.100 c.i. gutter for angle. & Ditto. for angle w. nozzle outlet for 75 downpipe, inc. cop. wire balloon gtg. & Ditto. for s.e.	Then follow with enumerated fittings, measured as extra over the gutter (SMM R10.11.2.1.0). Note use of single item for an angle incorporating a nozzle outlet and a copper wire balloon grating, and method of grouping different items with similar quantities. Balloon gratings can be enumerated as a pipework ancillary in accordance with SMM R10.6.4.1.0. Alternatively they can be included in the description of the enumerated item to which they relate (SMM R10.M4).
		<u>downpipes</u> 4·500 <u>add</u> sn. 150 4·650	Note the use of sub-headings to act as signposts and help in breaking down the dimensions.
2/	4·65	Pipe. st., nom. size: 75 dia., c.i. s&s med., w. ears cast on to BS460, jtd. in r.l. putty & fxg. 50 clear of wall w. gbdp. & secured w. rose hdd. g.i. nails & hwd. plugs let into bwk. at 2 m ccs. & Paintg. services, isoltd. surfs., ext. met., gth. ≤ 300, prep. & ③.	Downpipes are measured in metres over all fittings and descriptions are to include the type, nominal size, method of jointing, and type, spacing and method of fixing of supports and background, as SMM R10.1.1.1.1. Coating of pipes is covered by BS460 and does not therefore require specific mention. Painting to pipes, classified as services, is also measured in metres as ≤ 300 mm girth (SMM M60.9.0.2.0). This item is deemed to include work to saddles, pipe hooks, holderbats, and the like (SMM M60.C8). External painting must be so described (SMM M60.D1).
2/	1	E.o. c.i. pipe > 65 dia., for s.n. w. 225 proj.	The swanneck forms an enumerated item taken as extra over the pipe as SMM R10.2.4.5.0. If the pipe is ≤ 65 mm diameter, the type of fitting would not be stated, merely the number of ends. Jointing is described only when it is different from the pipe in which it occurs. No painting is measured to the pipe fittings as the additional length has already been included in the pipework.

PITCHED ROOF (Contd.)

Pipework ancillaries, such as gullies, are deemed to include cutting pipes and jointing (SMM R10.C8).
Note: where rainwater heads are to be provided, these are enumerated giving the type, nominal size, type of pipe, number and method of fixing any supports and the background (SMM R10.6.3.1.1).
Painting to rainwater heads is classified as painting services (SMM M60.D13).

ADJUSTMENT OF ROOFWORK FOR CHIMNEY STACK — EXAMPLE XIII

Timesing	Dimensions	Description	Notes
		Constn. Rafters	
	1.17	Ddt Rf. membrs., pitched 50 x 100 sn. swd.	Single rafter to be deducted across the chimney stack - the length is scaled between the centre lines of the trimmers.
2/	3.50	Ddt Ditto. 50 x 100 sn. swd. (trimg. rafters & Add Ditto. 75 x 100 sn. swd.	Substitution of thicker trimming timbers for those previously measured. There is no requirement to describe them as trimming timbers.
		len. 665 add spaces 2/50 100 trimmg. rafters 2/75 150 t. tenons 2/150 300 1.215	Build-up of length of trimmers including tusk tenons.
2/	1.22	Add Ditto. 75 x 100 sn. swd. (trimmers	

Note: An exactly similar procedure will follow for the adjustment of the ceiling joists; the dimensions are not included in this example as it is largely repetitive.

Timesing	Dimensions	Description	Notes
			Deduction of tiling and measurement of extra work to boundaries of stack (abutments and eaves) only operate when the void > 1.00 m² (SMM H60. M1-2). In this case the area is 1.25 x 0.67 = 0.84 m², and so no adjustments are made.
		Carpentry	Coverings are deemed to include work in forming voids ≤ 1.00 m², other than holes (SMM H60. C1b).
	0.67	Gutter bd., 225 x 25, sn. swd., trav. for lead. (gutter bott.	Gutter boarding is so classified and measured in metres giving an overall dimensioned cross-section description (SMM G20.14.3.2.0) When > 300 mm wide, it is measured in m² giving a dimensioned description.
3/	0.37	Rf. membrs., pitchd., 40 x 50 sn. swd. (brrs.	These bearers are measured in metres giving a dimensioned description similar to plates (SMM G20. 8.0.1.0).
		13.1	

1271.1

ADJUSTMENT OF ROOFWORK FOR CHIMNEY STACK (Contd.)

	0.93	Gutter bd., 300 x 25, sn. swd., trav. for lead. (lier bd.	Length of lier board provided up the roof slope to support the lead (0.93 m length needed to bridge the trimming rafters). Included in gutter board classification as side to gutter (SMM G20.M3 and D9).
2/	1	Gusset end to chy. gutter, irreg. shape, 225 x 225 x 25%a,- sn. swd.	Enumerated item as SMM G20.17. 0.1.0. giving a dimensioned description.
		Leadwk. gutter len. 665 add passgs. (ends) 2/150 300 965 width upstd. to wall 500 gutter bott. 225 up rf. slope 300 1.025	Note build-up of dimensions of lead lining to gutter, calculated where appropriate in accordance with the table of allowances given in SMM H.71. M2. The 500 mm upstand allowance in SMM 7 is high, as the more usual dimension is 150 mm.
	0.97	Lead gutter, o/a gth. 1.025 m, irreg. shape w. slopg. side, Nr. 5 lead to BS 1178, cop. nailed @ 50 ccs. to swd. on top edge.	Linear item in accordance with SMM H71.19.1.0.1-3, including a dimensioned description and stating the planes in which it is laid, and the type and quality of material and method of fixing. The work is deemed to include work to falls, dressing/wedging into grooves, hollows, recesses and the like (SMM H71.C1b and e).
		flashg. 665 add passgs. (ends) 2/100 200 865	
	0.87	Lead flashg., gth: 150, vert., Nr. 5 milled lead a.b. & fxg. w. lead tacks & lead wedgg. into prepd. grve. at top edge.	Linear item giving the girth and including the tacks and lead wedging (method of fixing) as SMM H71.53. Note the use of a.b. to avoid repeating the reference to the British Standard.
	0.67	Ptg. in flashgs. to bwk. in c.m (1:3).	Pointing to flashings is a separate item being in a different work section (SMM F30.6).
		apron 665 add passgs. (ends) 2/100 200 865	Passings of flashings are deducted in arriving at the length.
	0.87	Lead apron, gth: 300, vert. & slopg., Nr. 5 milled lead a.b. & fxg. w. lead tacks & lead wedgg. into prepd. grve at top edge.	Measured as SMM H71.11.1.0.2-3. Fixing is included in the description in accordance with SMM H71.53.

13.2

ADJUSTMENT OF ROOFWORK FOR CHIMNEY STACK (Contd.)

Timesing	Dimensions	Squaring	Description	Side notes
	0.67		Ptg. in flashgs. to bwk in c.m. (1:3).	Pointing to flashings is deemed to include cutting or forming grooves or chases (SMM F30.C4).
			stepped flashgs. 1.000 add passgs. (ends) 2/100 200 1.200	Length on slope scaled from Section.
2/	1.20		Lead flashg., gth: 200, stepped, Nr. 5 milled lead a.b. & fxg. w. lead tacks & lead wedgg. into prepd. grve. at top edge.	Stepped flashings are measured in accordance with SMM H71.10.1.0.4.
2/	1.00		Ptg. in flashgs. to bwk. in c.m. (1:3).	Linear item as SMM F30.6.
			soakers 100) 1.250 13 len. of soaker = lap + gauge + 25 = 65 + 100 + 25 = 190	Length of slope is divided by gauge of tiling to give the number of soakers on each side of the stack. The width of soakers is normally 150 mm. 50 ∟ 100
2/13/	1		Lead soakers, rt. ∠d., 190 lg. x 50 vert. x 100 hor., Nr. 4 milled lead a.b, handed to others for fxg.	Enumerated item for soakers with a dimensioned description as SMM H71.26.1.0.1.
			&	
			Lead soakers, a.b., fixg. only.	Separate enumerated fixing only item for soakers as SMM H60. 10.5.1.1.

13.3

ASPHALT FLAT ROOF

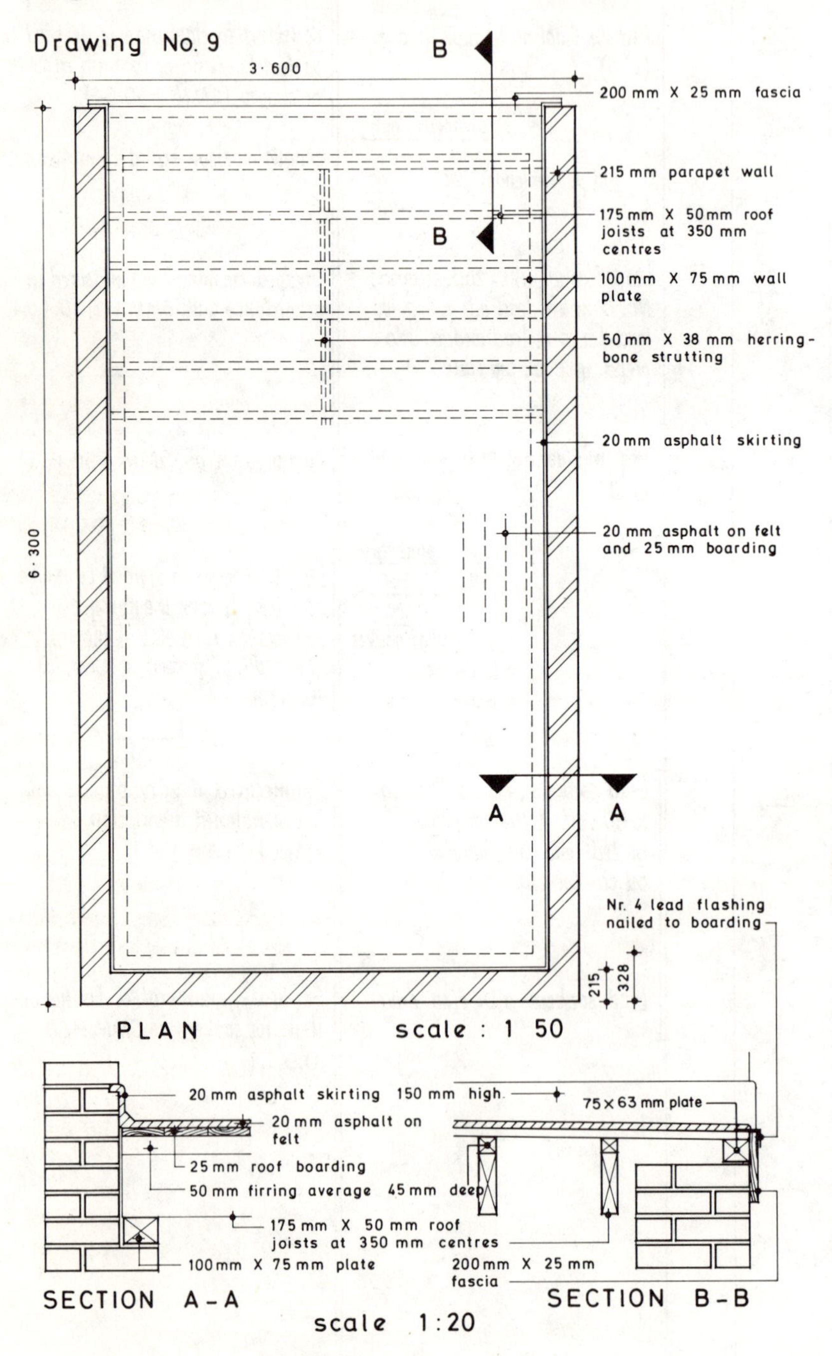

ASPHALT FLAT ROOF — EXAMPLE XIV

Timesing	Dimension	Description	Notes
		Constn. len. of plate: 6.300 less 328, 215 = 543; 5.757	Build up of length of plate in waste.
2/	5.76	Plate 100 x 75 sn. swd	Classified as SMM G20.8.0.1.0, with a dimensioned description.
		3.600 less 2/215 = 430; 3.170	
	3.17	Plate 75 x 63 sn. swd.	
		5.757 less ends 2/50 = 100; 350) 5.657 = 17+1	Calculation in waste to arrive at the number of flat roof joists – the adjusted roof length is divided by the spacing of the joists to give the number of spaces, to which one is added to give the number of joists.
18/	3.17	Rf. membrs., flat, 50 x 175 sn. swd.	Flat roof joists are classified as roof members, flat, as SMM G20.9.1.1.0.
		&	
		Individual suppts., 50 x 45 (av.), diff. cross-sectn. shapes, sn. swd. (firrgs.	Firring pieces are classified as individual supports and measured in metres giving a dimensioned cross-section description as SMM G20.13.0.1.1 and G20.D7.
	5.76	Joist struttg., 50 x 38 sn. swd., herrgbone, 175 dp. jsts.	Herringbone strutting is measured over the joists in metres, stating the depth of joists (SMM G20.10.1.1.0 and G20.M1). With asphalt work it is necessary to supply a plan of each level indicating the extent of the work and its height above ground level, together with restrictions on the siting of plant and materials (SMM J21.P1a). Note the build-up of the length and width of the asphalt in waste.
		Covergs. len.: 6.300 less ppt. wall 215 = 6.085; add fascia 25 = 6.110. width: 3.600 less 2/215 = 430; 3.170	

14.1

ASPHALT FLAT ROOF (Contd.)

Timesing	Dimensions	Squaring	Description	Explanatory notes
	6·11 3·17		Mastic asp. roofg., width > 300, pitch of 1 in 80, to BS 1162, 20th. in 2 cts. on single layer of rfg. felt to BS 747 (type 2a), nailed to bdg. (m/s).	The asphalt is measured in m², stating the kind and quality of materials, including underlays, and thickness and number of coats and nature of base on which applied (SMM J21.S1-3). The description includes the pitch and width range as SMM J21.3.4.1.0. Any cutting, notching, bending and lapping the underlay is deemed to be included (SMM J21.C1b). Any surface treatment such as chippings or tiles is included in the description (SMM J21.S4).
	6·09 3·17		Rf. bdg., width > 300, 25 sn. swd. nailed to swd. jsts. (m/s).	Roof boarding is measured in m², giving the width range and dimensioned description (SMM K20.4.1.1.0). It is only classified as sloping if > 10° from the horizontal. All labours are deemed to be included (SMM K20.C1a). It is, however, necessary to state the nature of the background and method of jointing or form of construction where not at the discretion of the contractor (SMM K20.S2-3).
	3·17		Mastic asp. sktg. gth. ≤ 150, 20 th. in 2 cts.	Skirting is measured in metres giving the girth range as SMM J21.5.1.0.0. Asphalt skirtings, fascias and aprons are deemed to include edges, drips, arrises, internal angle fillets, dressing over tilting fillets, turning nibs into grooves, including extra material, angles and stopped and fair ends (SMM J21.C4).
			len. 2/6·085 = 12·170	
	12·17		Ditto. gth. 150 – 225, do.	The top of the skirting will follow an enlarged horizontal brick joint while the bottom will follow the slope of the roof, so that the face width varies.
2/	0·15		(returns	
	3·17		Rdd. edge to asp.	Measured over fascia in metres as SMM J21.14.

14.2

ASPHALT FLAT ROOF (Contd.)

		Description	Notes
		3·600 less 2/75 150 3·450	
3·45		Fascia bd. 25 x 200 wrot swd. & Paintg. gen. isoltd. surfs., ext. wd., girth. ≤ 300, k.p.s. & (3).	Linear item as SMM G20.15.3.2, with a dimensioned overall cross-section description. Painting to fascia boards is described as to general isolated surfaces and is measured in metres as ≤ 300 mm girth (SMM M60.1.0.2.0). External work must be so described (SMM M60. D1), and the nature of the base and the painting system must be stated (SMM M60. S1-8). The sequence of the description is likely to appear unusual, but follows closely the order of items in SMM7 and is well suited to project specification cross-referencing and computerisation.
3·17		Lead flashg., gth: 125, vert., Nr. 5 lead to BS 1178, welted & cop. nailed @ 50 ccs. to swd. on top edge.	Lead flashings are measured in metres, giving a dimensioned description as SMM H71. 10.1.0.3. The description is to include the type and quality of materials, method of fixing and type of support materials (SMM H71. S1-6). It is assumed that rainwater goods are measured elsewhere. Many flat roofs are covered with built-up felt, which is measured in m² giving the particulars listed in SMM J41.2.1 and J41.S1-3. There are also a variety of linear items, such as abutments, eaves, skirtings, flashings, aprons and gutters, for which the girth is stated (SMM J41.3-13.2).

14.3

LEAD FLAT ROOF Drawing No. 10

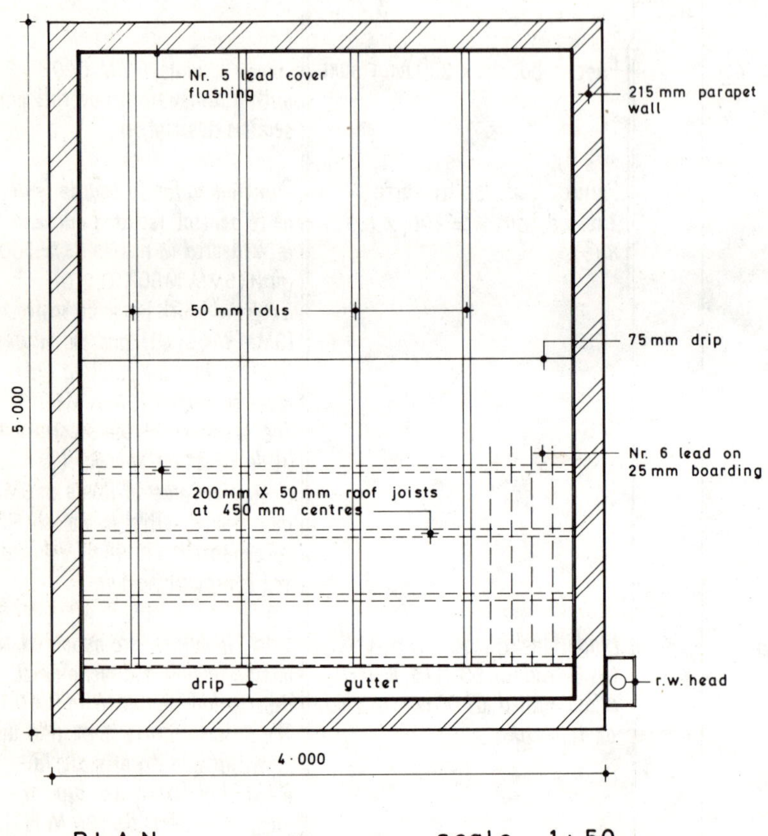

PLAN scale 1 : 50

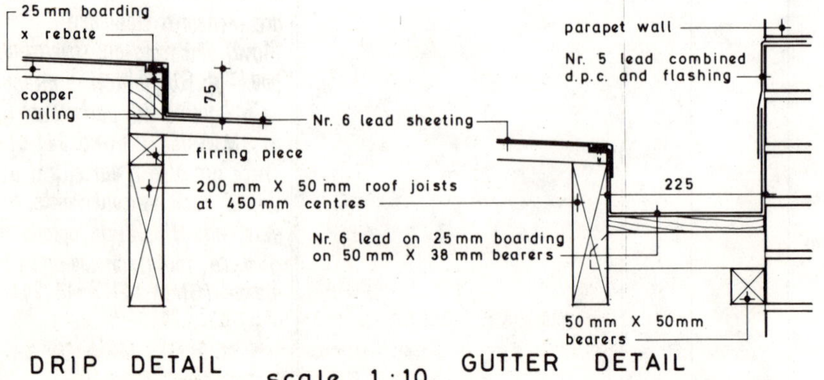

DRIP DETAIL scale 1 : 10 GUTTER DETAIL

LEAD FLAT ROOF — EXAMPLE XV

Timesing	Dimensions	Description	Notes
		Constn. 5·000 less walls 2/215 430 gutter 225 one end space 50 other end to ℄ of jst. 25 730 450)4·270 10+1	If the spacing between the centres of joists is not to exceed 450 mm, then 11 joists are required.
		len. 4·000 less walls 2/215 430 3·570 add. beargs. 2/100 200 3·770	
11/	3·77	Rf. membrs., flat, 50 x 200 sn. swd. (rf. jsts.	Classification as SMM G20.9.1.1.0.
	3·57	Plate 50 x 50 sn. swd. plugd. to bwk. (brr.	Bearers are classified as plates and measured in metres, giving a dimensioned description (SMM G20.8.0.1.0). Fixing is described if not at the discretion of the contractor (SMM G20.S2).
11/	2	2ce solignum end of flat rf. membr. b.f.	Enumerated item for treating ends of roof joists with preservative. If measured under work section M60 (painting), the description could read '2ce solignum, general isolated surfaces, external wood area ≤ 0·50 m²'
		len. 4·270 add end spaces 75 4·345 add drip 50 4·395	
	4·40 3·57	Rf. bdg., width > 300, 25 sn. swd. t.&g. nailed to swd. jsts. (m/s).	Particulars as SMM K20.4.1.1.0 measured in m², and including nature of background (SMM K20.S3).
		Firrg. pieces av. depth lower end: 0, 40, 2)40, av. 20 upper end: 120, 160, 2)280, 140 20 2)160 av. for whole rf. 80	Build-up of average depth of firrings over whole area of roof.
11/	3·57	Individual suppts., 50 x 80 (av.) varying cross-section shapes, sn. swd. (firrgs.	The firrings are classified as individual supports, giving a dimensioned overall cross-section description (SMM G20.13.0.1.1 and G20.D7).

15.1

LEAD FLAT ROOF (Contd.)

Timesing	Dimensions	Squaring	Description	Notes
4/	4.35		Individual suppts., 30 dia. x 50 hi., irreg. shaped area w. semi-circ. top, wrot swd. (rolls	Rolls will require a wrot finish to receive the lead. They are measured in metres as individual supports, giving a dimensioned overall cross-section description and including reference to the irregular area (SMM G20.13.0.1.4 and G20.D7).
	3.57		Individual suppt. 50 x 75, sn. swd. (drip & Gutter bd., 225 x 25 sn. swd.	Drips are measured in metres as individual supports, giving a dimensioned overall cross-section description (SMM G20.13.0.1.0). Cross rebates have not been measured as the work is deemed to include labours on items of timber (SMM G20.C1). There is no gutter side in this example; gutter boards and sides are each measured overall as individual components, in metres if width is ≤ 300 mm, but grouped together (SMM G20.M3).
			450) 3.570 8+1	
9/	0.30		Plate 38 x 50 sn. swd. (brrs.	Bearers are classified as plates and measured in metres with a dimensioned description (SMM G20.8.0.1.0).
			Covergs. width 3.570 add rolls 4/250 1.000 upstand 2/500 1.000 5.570 len. 4.345 add upstand 500 turndown (gutter) 50 75 drip 230 5.125	Allowances for rolls and upstands are made in accordance with the table of allowances in SMM H71.M2. The allowance of 180 mm for a 50 mm drip needs increasing by 2/25 mm (two extended vertical lengths of lead) to give an allowance of 230 mm for a 75 mm deep drip.
	5.57 5.13		Lead sheet rf. coverg., pitch: 1 in 54, Nr. 6 milled lead to BS 1178, cop. nailed to swd. bdg. @ 50 ccs. on 2 edges of ea. sheet.	Lead sheet flat roof coverings are measured in m², giving the pitch as SMM H71.1.1, and type and quality of materials, and method of fixing and type of support as SMM H71.S1-6. In practice, some of the more detailed information such as copper nailing could be condensed by inserting the appropriate reference to the project specification, where available.
			gutter len. 3.570 add upstand 500 50 drip 180 thro'. wall 300 4.550 width bottom 225 upstds. 2/500 1.000 1.225	The upstand/downstand allowance in SMM7 does seem very high, but it has, nevertheless, been adopted.

15.2

LEAD FLAT ROOF (Contd.)

Timesing	Dimension	Description	Side notes
	4.55	Lead sheet covd. gutter, o/a gth: 1.225 m, U shaped, Nr. 6 milled lead a.b.	Lead gutters are measured in metres, giving a dimensioned description and classified as SMM H71.19.1.0.0.
		len. less 5.000 — 4.570 walls 2/215 430 — 3.570 4.570 — 2/8.140 16.280 add cornrs. 4/215 860 17.140	Leadwork is deemed to include laps, seams, ends, angles and intersections, rolls, upstands and downstands, and dressing/wedging into grooves, hollows and recesses (SMM H71.C2). It will, however, be necessary to allow for the length of the dpc measured on its centre line. Note build up of length of flashing/dpc in waste.
	17.14	Lead sheet comb. flashg./dpc, gth: 365, hor., vert. & stepped Nr. 5 milled lead a.b. & fxg. w. lead tacks @ 750 ccs.	Flashings are measured in metres, giving a dimensioned description as SMM H71.10.1.0.1,3,4. The dpc being an integral part of the component is included in the description, and also the method of fixing (SMM H71.S3).
1/2/	0.22	Lead sheet retn., gth: 215, vert., Nr. 5 milled lead a.b. (steps & gutter drip	The level of the dpc will have to be varied in the vicinity of the drips to accommodate the difference in roof and gutter levels (SMM H71.18.1.0.3).
			No additional items are required for dressing lead around the rainwater outlet and into the rainwater head, as they are deemed to be included in the measured work (SMM H71.C2e). Rainwater goods are assumed to be measured elsewhere. Drawings are needed to show the extent of the roofing work and its height above ground level, including the location and spacing of all laps, drips, welts, cross welts, beads, seams, rolls, upstands and downstands (SMM H71.P1a and b).

15.3

9 Measurement of Internal Finishings

SEQUENCE OF MEASUREMENT

In the measurement of this class of work it is essential to adopt a logical sequence in the taking off work. Where the work is extensive it is advisable to prepare a schedule of finishings in the form illustrated on pages 136 and 137 unless supplied by the architect. This will considerably simplify the taking off and reduce the liability to error.

This section normally includes ceiling, wall and floor finishings and the principles of measurement of most of these items of work are covered in a variety of work sections ranging from M10 to M60 of the *Standard Method*, but with timber board flooring in K20 and skirtings and similar members in P20. As indicated in chapter 7, floor finishings may be taken off when measuring the floors or be left to be taken later with internal finishings. On balance, the best arrangement is probably to take hollow floor finishings (boarded flooring) with the floors, since they form an integral part of the floor construction and the work is related to an associated trade. Finishings to solid floors which can be of infinite variety are best measured with internal finishings.

The taker-off should work systematically through the building, room by room, preferably working from top to bottom, and recording the details of finishings in a schedule in the majority of cases. It will probably prove helpful to number the rooms for this purpose and to mark each room in some distinctive way once the details have been extracted. In the case of a very small building with only a limited variety of finishings the details can often be transferred direct to the dimensions paper, without the need for a schedule, but the same systematic approach should be adopted.

The order of measurement of internal finishings on each floor will normally be: (1) ceilings, (2) walls and (3) floors. The measurement of the main areas of wall finishings will be followed by linear items, such as cornices, picturerails, dado rails and skirtings, and working in this sequence (top to bottom).

Work to walls, ceilings and floors in staircase areas and, on occasions, to plant rooms shall be given separately.

CEILING FINISHINGS

The ceiling area is measured between wall surfaces with the area of each type of finish measured separately in m^2, followed by any associated labours, such as arrises to beams. For instance, with a plaster ceiling, covered in the *Standard Method* under the work section M20, the various particulars listed in SMM M20.S1–6, such as kind, quality, composition and mix of materials, method of application, nature of surface treatment and nature of base are to be given. In addition, the thickness and number of coats are to be stated. Work to ceilings and beams over 3.50 m above floor level (measured to ceiling level in both cases), except in staircase areas, shall be so described, stating the height in further stages of 1.50 m, that is $>$ 3.50 mm and $\leq$ 5.00 m, $>$ 5.00 m and $\leq$ 6.50 m, and so on (SMM M20.M4).

Plasterboard ceilings are dealt with similarly, stating the thickness of the plasterboard. Skim coats of plaster to plasterboard ceilings are included in the same items.

It will be noted that work in staircase areas is given separately to enable the estimator to price for working off staircase flights and/or in restricted areas. Work to sides and soffits of attached beams and openings is classified as work to abutting ceilings (SMM M20.D5).

WALL FINISHINGS

The measurement of wall finishings is taken from floor to ceiling, including the work behind wood skirtings and similar features, disregarding the grounds (SMM M20.M2).

The girth of each room is usually built up in waste and the total girth of rooms of the same height and finishing transferred to the dimension column. The method of measurement varies with the type of finishing, and reference needs to be made to the appropriate work section, such as M20 for renderings and plastered coatings, M40 for quarry and ceramic tiling and M41 for terrazzo tiling. The measurement of the main area of wall finishing of each type is followed by its associated linear items.

Plaster to walls and isolated columns in widths $\leq$ 300 mm is measured in metres, and no deduction shall be made for voids $\leq$ 0.50 m^2 (SMM M20.M2). Work to sides and soffits of openings and sides of attached columns shall be regarded as work to the abutting walls.

Rounded angles and intersections to plaster in the range 10–100 mm radius are measured in metres (SMM M20.16 and M20.M7), while those of smaller radius are included in the plasterwork rates.

Working plaster over and around obstructions, pipes and the like, and into recesses and shaped inserts is deemed to be included (SMM M20.C1b).

SCHEDULE OF INTERNAL FINISHINGS

Room	*Ceiling Finishing*	*Decorations to Ceiling*	*Wall Finishing*	*Decorations to Walls*	*Cornice or Cove*
Hall	10 mm gypsum plaster-board to BS 1230 type b and 3 mm coat of gypsum plaster	Prepare and twice emulsion paint	Two coats gypsum plaster to BS 1191, class B, finishing 13 mm thick	Prepare and twice emulsion paint	150 mm plaster cornice
Lounge	Ditto	Ditto	Ditto	Ditto	Ditto
Kitchen	Ditto	Prepare and two coats of oil paint	Cream glazed ceramic tiling 1.00 m high on cement and sand backing and plaster as before above	Prepare and two coats of oil paint above tiling	–
Bedroom 1	Ditto	Prepare and twice emulsion paint	Two coats gypsum plaster, to BS 1191, class B	Prepare and twice emulsion paint	50 mm plaster cove
Bedroom 2	Ditto	Ditto	Ditto	Ditto	Ditto
Bathroom	Ditto	Prepare and two coats of oil paint	Cream glazed ceramic tiling 1.00 m high on cement and sand backing and plaster as before above	Prepare and two coats of oil paint above tiling	–
W.C.	Ditto	Ditto	Ditto	Ditto	–
Out-building	5 mm coat of gypsum plaster to concrete soffit	Clearcolle and twice whiten	Fair-faced brickwork	Prepare and twice emulsion paint	

Frieze	*Picture rail*	*Skirting*	*Floor Finishing*	*Any Other Features*
450 mm prepare and twice emulsion paint	25 × 50 mm moulded softwood plugged to brickwork	25 × 125 mm moulded softwood plugged to brickwork	Oak blocks 25 mm thick, laid herring-bone on 25 mm screed	20 mm rounded arrises to walls
Ditto	Ditto	Ditto	Ditto	Ditto
–	–	16 × 152 mm red quarry tile on cement and sand backing	152 × 152 × 16 mm red quarry tiles on 9 mm bedding & 25 mm screed	–
–	–	25 × 100 mm moulded softwood plugged to brickwork	229 × 229 × 3 mm thermo-plastic floorg. tiles to BS 2592 on 47 mm screed	–
–	–	Ditto	Ditto	–
–	–	25 mm hardwood quadrant plugged to brickwork	5 mm sheet rubber flooring to BS 1171 on 45 mm screed	–
–	–	Ditto	Ditto	–
–	–	–	40 mm granolithic paving	–

Cornices, mouldings and coves are measured in metres (length in contact with base) stating the girth or giving a dimensioned description (SMM M20.17–19.0.1–2.0). Ends, internal angles, external angles and intersections are each enumerated extra over the appropriate linear items, giving adequate details as SMM M20.23.1–4.1.0.

Where the wall finishing is of tiles, slabs or blocks of cast concrete, precast terrazzo, natural stone, quarry or ceramic tiles, plastics, mosaic or cork, full particulars of the finishings are to be given, such as kind, quality and size of materials, nature of base, surface treatment, method of fixing, and treatment and layout of joints (SMM 40/41/50.S1–8). Work > 300 mm in width is measured in m^2, while narrower widths are measured in metres.

In situ finishings are measured in a similar manner to plastering.

SKIRTINGS AND PICTURE RAILS

Timber skirtings, picture rails, dado rails, architraves and the like are measured in metres, giving a dimensioned overall cross-section description as SMM P20.1.1.0.1–4. The work is deemed to include ends, angles, mitres, intersections and the like, except on hardwood items > 0.003 m^2 in sectional area (SMM P20.C1). The description is to include the kind and quality of timber and whether sawn or wrought and method of fixing where not at the discretion of the contractor (SMM P20.S1–9).

In situ and tile, slab and block skirtings are measured in metres, stating the height or height and width as appropriate, and are deemed to include fair edges, rounded edges, ends, angles and ramps (SMM M40.12.1–2 and M40.C8).

FLOOR FINISHINGS

Floor finishings are measured in m^2, irrespective of their width, and are classified under three categories according to slope (SMM M40.5.1–3), and where floors are laid in bays, the average size of bay is stated. Screeds are measured in an identical manner to floor finishings. Dividing strips at door openings and the like, between different types of floor finishing, are measured in metres giving a dimensioned description (SMM M40.16.4.1.0).

PAINTING AND DECORATIONS

The painting and decorating of ceilings, cornices and walls are classified as to general surfaces and measured in m^2, except for work on isolated surfaces ≤ 300 mm in girth which is given in metres, or work in isolated areas ≤ 0.50 m^2 which is enumerated (SMM M60.1.0.1–3.0). Multi-coloured work is separately classified and is defined as the application of more than one colour on an individual surface, except on walls and piers or on ceilings and beams (SMM M60.D2). Paintwork is deemed to include rubbing down with glass, emery or sand paper (SMM M60.C1). Full particulars of painting, decorating and polishing shall be given in accordance with SMM M60.S1–8.

Work in staircase areas and plant rooms is to be kept separate because of the extra costs involved (SMM M60.M1). Work to ceilings and beams over 3.50 m above floor level (measured to ceiling level in both cases), except in staircase areas, shall be so described stating the height in further 1.50 m stages (SMM M60.M4).

The supply and hanging of decorative papers and fabrics is separated between walls and columns; and ceilings and beams; with areas >0.50 m^2 measured in m^2 and those ≤ 0.50 m^2 enumerated. Where these items include raking and curved cutting and/or lining paper, these are included in the description (SMM M52.1–2.1–2.0.1–2). Border strips are measured in metres, including cutting them to profile in the description where appropriate (SMM M52.3.0.0.1). Material particulars are to include the kind and quality of materials, including the manufacturer and pattern (SMM M52.S1), although these particulars may be included in a preamble clause or covered by a project specification reference. The *Code of Procedure for Measurement of Building Works* states that the width of rolls and type of pattern would need to be given before wallpaper could be considered fully described.

WORKED EXAMPLE

A worked example follows giving the dimensions of the ceiling, wall and floor finishings to a small building.

INTERNAL FINISHINGS Drawing No. 11

SCHEDULE OF FINISHINGS			
LOCATION	**CEILING**	**WALLS**	**FLOOR**
ENTRANCE PASSAGE	10 mm plasterboard skimmed with 3 mm coat of hardwall plaster and twice emulsion painted	render in cement and sand (1:3) 10 mm thick and set with 3 mm hardwall plaster and finished with 2 undercoats and 1 coat of hard gloss paint	3 mm thermoplastic tiles on 52 mm cement and sand screed
WAITING ROOM	DITTO	render and set as before: 450 mm frieze twice emulsion painted, remainder papered with pattern paper p.c. £ 5·00 p per piece and frieze border p.c. £ 0·40 p per metre	30 mm thick oak block flooring laid herring bone, with two block plain border on 25 mm cement and sand screed
OFFICE	DITTO	render and set as before and twice emulsion paint	DITTO
NOTE: All rooms are 2·400 high (20 mm coved internal angles)			

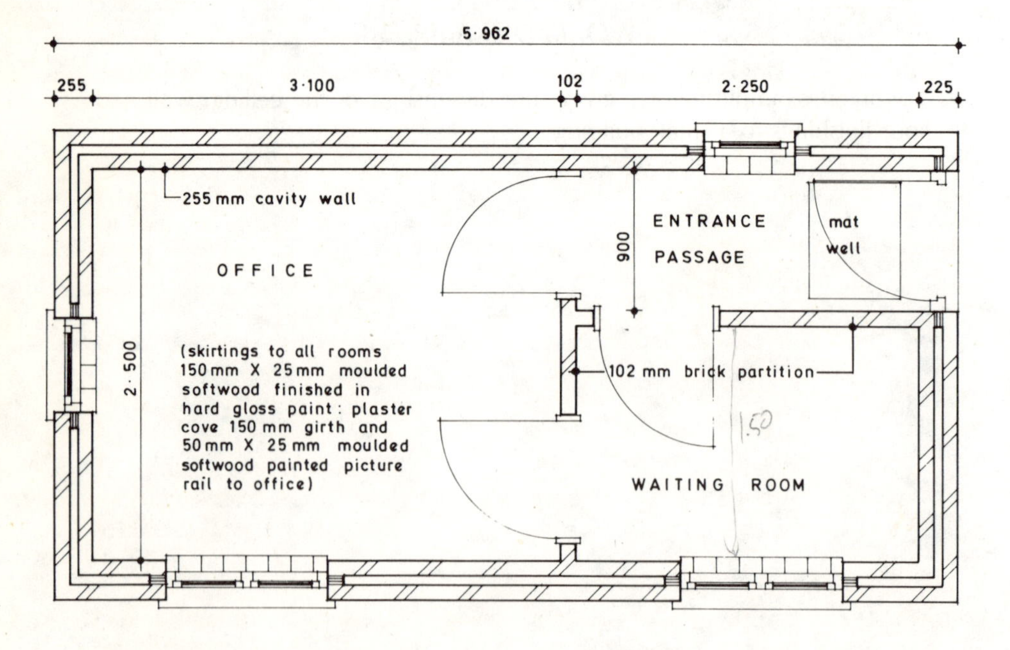

PLAN scale 1:50

INTERNAL FINISHINGS

			Ceilgs.
		less 900	2·500
		102	1·002
		width of W. Rm.	1·498
	3·10 2·50	Pla. clgs. gypsum lath. pla. bd. 10 th. to BS 1230 fxd. w. 32 galvd. rd. hdd. nails to u/s of swd. jsts., scriming. jts., & skim ct. of nt. thistle bd. fin. gypsum pla. to BS 1191 class B, 3 th, w. trowld. fin.	(Office
	2·25 1·50		(W. Rm.
		&	
	2·25 0·90	Paintg. gen. surfs., seal & 2 ^ce^ emulsn. pt. pla. clgs.	(Ent. Pass.
			Walls
		Office	3·100
			2·500
		2/	5·600
			11·200

16.1

EXAMPLE XVI

The work is deemed to be internal unless described as external (SMM M20. D1).

Adopt a logical sequence of taking off, such as ceilings, walls and floors.

Plasterboard to ceilings is measured in m^2 giving the thickness of plasterboard and the thickness and number of skim coats as SMM M20.2.1.2.0, and appropriate particulars listed in SMM M20. S1–6.

Alternatively the type of plasterboard and associated particulars could be covered in a preamble clause or by cross reference to a project specification.

Work to ceilings over 3·50 m above floor are so described in 1·50 m stages, except in staircase areas (SMM M20.M4).

Plasterboard is deemed to include joint reinforcing scrim (SMM M20. C3).

See SMM M20. D5–6 for the measurement of finishes to beams and columns, and note the need to keep work to staircase areas and plant rooms separate (SMM M20.M3).

Decorations to ceilings are measured in m^2 and classified as to general surfaces (SMM M60.1.0.1.0), & particulars given in accordance with SMM M60.S1-8.

Painting and decorations to ceilings over 3·50 m above floor are so described in stages of 1·50 m, except in staircase areas, which are kept separate and so described (SMM M60. M4). Rubbing down is deemed to be included and does not require specific mention (SMM M60. C1).

Firstly, build up the girth of rooms in waste. The full height of the plaster is taken ignoring the grounds behind wood skirtings (SMM M20.M2).

INTERNAL FINISHINGS (Contd.)

Timesing	Dimensions	Squaring	Description	Side notes
			Walls W. Rm. 2.250 1.498 2/3.748 7.496 Ent. Pass. 2.250 900 2/3.150 6.300	All measurements are taken on the actual wall surfaces, not on the centre line of the plaster, and no deductions are made for voids ≤ 0.50 m² (SMM M20.M2).
	11.20 2.40 7.50 2.40 6.30 2.40		Pla. to walls, ct. & sd. (1:3), in 2cts., 10th. & gypsum finshg. ct. pla. type B to BS 1191 Part 2, 3th. to bwk., trowld. fin. (Office (W. Rm. (Ent. Pass.	Full particulars of plaster are given as required by SMM M20.1.1.1.0 and M20.S1-6, keeping the work to walls and ceilings separate. The kind, quality, composition and mix of materials, nature of base and surface treatment are included in the description, but they could alternatively be given in a preamble clause or a cross reference made to a project specification. Follow with any associated labours which require separate measurement.
			2.400 less sktg. 150 2.250	
3/4/	2.25		Rdd. ∠ to pla., rad. 10 – 100.	Rounded angles and intersections are measured in metres finishing at the top of skirtings, where the radius is in the 10 to 100 mm range (SMM M20.16 and M20.M7), and those ≤ 10 mm radius do not require measuring. When measuring the work to doors adjustment will be made to rounded angles, where the plaster finishes against the door frame or lining.
	11.20		Pla. cove, gth: 150, gypsum pla. class B., to BS 1191 Part 1, to bk. & pla. bd. surfs., fltd. fin. (Office	Coves and cornices are measured in metres, stating the girth or giving a dimensioned description as SMM M20.17.0.1.0, with ends, internal angles, external angles and intersections enumerated as extra over the component as SMM M20.23.1–4.1.0.
4/	1		Ex. for int. ∠ to pla. cove. (do.	Precast plaster components are measured in accordance with SMM M20.25.1.0.0.
	11.20 0.10		Ddt. Pla. to walls a.b. (Office	Regardless of the area involved the wall and ceiling finishes are deductable as the cove is on the boundary of each measured area and does not therefore constitute a void (SMM General Rules 3.4).
4/	0.10		Ddt. Rdd. ∠ to pla., rad. 10 – 100. (do.	

INTERNAL FINISHINGS (Contd.)

Dimensions	Description	Notes
		However it does not appear justifiable to deduct the skim coat on the plasterboard ceiling for the width occupied by the cove as it forms a small part of a combined item, and a balanced and logical view is needed, especially having regard to the double adjustment that would be required.
	Decs. to walls	
11·20 2·25	Paintg. gen. surfs., seal & 2ce emulsn. pt. pla. walls. (Office	Decorations are taken from ceiling level to top of skirting leaving the cove to be adjusted later.
6·30 2·25	Paintg. gen. surfs., seal & ② & ① hd. gloss pt. on (Ent. Pass. pla. walls. ht. 2·400 less sktg. 150, frieze 450 = 600 1·800	Note: all adjustments of internal finishings for window and door openings will be taken with the windows and doors. Painting particulars are listed in SMM M60.S1-8. ② represents two undercoats and ① one finishing coat.
7·50 1·80	Decorative paper to walls, rub down, size & hang patt. paper w. cellulose adhesive on pla. (W. Rm	Paperhanging is measured in m^2, giving the particulars listed in SMM M52.S1-4, and classified as to position (SMM M52.1.1.0.0). Details of the kind and quality of materials, manufacturer and pattern could be given in a preamble clause or the project specification.
	Provide the P.C. Sum of £30 for the supply of wall paper. & Add for profit	The supply is covered by a separate prime cost item, with provision for the addition of contractor's profit. This avoids the need for the estimator to obtain quotations and permits the choice of paper to be delayed. However, the choice of pattern can influence the number of rolls required and the labour needed to hang the paper.
7·50	Border strip fxd. w. cellulose adhesive to pla. walls (supply m/s). (W. Rm.	Border strips are measured in metres as SMM M52.3 as regards hanging with a separate prime cost supply item (in previous item). Border strips are deemed to include mitres and intersections (SMM M52.C2).
7·50 0·45	Paintg. gen. surfs., seal & 2ce emulsn. pt. pla. clgs. (frieze W. Rm. 11·200 less crnrs. 4/100 = 400 10·800	The decorations to friezes are added to similar work on abutting walls or ceilings.
	Cove	
10·80 0·15	Paintg. gen. surfs., seal & 2ce emulsn. pt. pla. clgs.	In this case a similar type of work is carried out on the ceiling, and the painting of the cove has therefore been included in with the decorations to the ceiling.

16.3

INTERNAL FINISHINGS (Contd.)

	10·80 0·10	Ddt. ditto.	Deductions of areas of painting previously taken on ceilings and walls for the area occupied by the cove.
	11·20 0·10	Ddt. ditto. pla. walls.	
		Pic. rl.	
	11·20	Pic. rl. 25 x 50 wrot. swd. mo. fxd. w. grds. plugd. to bwk. (Office & Paintg. gen. isoltd. surfs., wd., gth, ≤ 300, k.p.s. & ② & ① hd. gloss.	Picture rails are measured in metres, giving a dimensioned overall cross-section description as SMM P20.1.1.0.0. The work is deemed to include ends angles, mitres, intersections and the like, except hardwood items > 0·003 m^2 sectional area. The method of fixing is described where not at the contractor's discretion (SMM P20.C1). Both picture rails and skirtings may be fixed without grounds, by plugging to brickwork and this needs to be specifically mentioned. Picture rails are grouped with skirtings when measuring paintwork as general isolated surfaces of wood, and if the girth is ≤ 300 mm, it is measured in metres (SMM M60.1.0.2.0).
		Sktgs.	
	11·20 7·50 6·30	Sktg. 25 x 150 wrot swd. mo., fxd. w. grds. plugd. to bwk. (Office (W. Rm. (Ent. Pass. & Paintg. gen. isoltd. surfs., wd., gth. ≤ 300, k.p.s. & ② & ① hd. gloss. dr. 750 add archves 2/50 100 850	Skirtings are measured similarly to picture rails. Tile skirtings are measured in metres stating the height and width (SMM M40.12.2.0.0) and ends and angles are deemed to be included (SMM M40.C8). Painting work is deemed to be internal unless described otherwise (SMM M60.D1). Note: ② and ① represents two undercoats and one finishing coat.
3/2/	0·85 0·90	Ddt. Sktg. 25 x 150 a.b. (dr. opgs. & Ddt. Paintg. gen. isoltd. surfs., wd., gth. ≤ 300, a.b.	Adjustment of skirtings at door openings can be taken at this stage or alternatively with the doors.
2/	0·10	Add last two items. (reveals to ent. dr. 16.4	

INTERNAL FINISHINGS (Contd.)

		Floor Finishgs.	Screeds are measured in m², giving the particulars listed in SMM M10.5.1 and M10.S1-6. The details include the kind, quantity, composition and mix of materials, and nature of surface treatment and base.
	3.10 2.50	Screed to flr., lev., 25 ct. & sd. (1:3), fltd. on conc. (Office	
		&	
	2.25 1.50	(W. Rm.	Alternatively, these could be given in a preamble clause or a project specification, with appropriate cross referencing.
	0.75 0.10	Wd. blk., flrg., lev., herrgbone pattn., 300 x 75 x 30 European (dr. opg. Oak wrot t.&g. blks. dipd. & jtd. in hot bit., on fltd. scrd. (m/s) w. 2 block plain margins a/rd., & seal, body in & 2cts. of wax pol.	The description of the wood blocks is to include the appropriate particulars listed in SMM M42.5.1.0.1 and M42.S1-8. Plain block borders or margins are included in the description of the floor finish. The SMM does not require floor finishes in door openings to be kept separate. The nature of the surface treatment, including wax polishing or resin sealing coat is inserted in the description of the wood blocks (SMM M42.S5).
	2.25 0.90	Screed to flr., lev., 52 ct. & sd. (1:3) trowld. (Ent. Pass to rec. thermp. tiles, on conc.	Detailed in accordance with SMM M10.5.1 and M10.S1-6.
2/	0.75 0.10	(int. dr. opgs.	
		&	
	0.90 0.10	Plastics flr., lev., (ext. dr. opg. 250 x 250 x 3 th. mottled thermp. tiles to BS 2592, fxd. w. adhesive on trowld. scrd. (m/s).	The description includes the appropriate particulars listed in SMM M50.5.1.1.0 and M50.S1-8.
2/	0.75	Dividing strip, 5 x 40 ebonite, bedded in c.m. (1:3).	Measured in metres between different floor finishes in accordance with SMM M50.13.5.1.0 giving a dimensioned description. No deduction of floor finishes is required for the mat well as the area of the void does not exceed 0.50 m² (See SMM M50.M1). Fair joints to the accessories are deemed to be included (SMM M50.C1a).
	1	Mat. fr. galv. m.s., size 750 x 600 o/a of 38 x 38 m.s. Ls, welded at cornrs. inc. settg. in ct. & sd. (1:3).	Enumerated item, giving a dimensioned description.

16.5

10 Measurement of Windows

ORDER OF MEASUREMENT

The normal order of measurement of windows is to take the windows first, followed by associated components, ironmongery, glazing and painting. This is followed by the adjustment of the window opening working in a logical sequence such as

(1) Deduction of walling and finish on both faces.
(2) Head of opening (arches and lintels).
(3) External reveals.
(4) Internal reveals.
(5) External sill.
(6) Internal sill.

By working systematically through the dimensions in this way, the risk of omission of any items is much reduced.

Some surveyors prefer to measure the adjustment of the openings in the first instance and then follow with the actual windows.

WINDOWS

Wood casements and sash windows together with their frames are enumerated and described, accompanied by a dimensioned diagram (SMM L10.1.0.1.0), giving the appropriate particulars listed in SMM L10.S1–10. Standard sections are identified (SMM L10.M1). Member sizes will normally be nominal sizes (SMM L10.D1). Standard metal windows are measured similarly. The work is deemed to include architraves, trims, sills, sub-frames and the like and finishes where they are part of the component, and also ironmongery and glazing where supplied with the component, and fixings and fastenings (SMM L10.C2b–e & g). The enumerated window item is followed by any additional items such as sub-sills, window boards, bedding and fixing frames, painting the backs of frames, and the like.

When measuring glazing, allowance must be made for the rebates in the enclosing members of the window and panes of irregular shape shall be measured as the smallest rectangular area from which they can be obtained and so described (SMM L40.1.1.1–2.2 and L40.M3. A full description of the glazing must be given in accordance with SMM L40.S1–5, and it is measured in m^2, except for louvres which are enumerated, giving the pane size classification as SMM L40.1.1.1–2. Where glazing rebates are 20 mm or more deep, they are classified in 10 mm stages.

Ironmongery is enumerated in accordance with SMM P21.1.1, comprising the supply and fixing. However, prime cost sums are often used to cover the supply of ironmongery, when separate items for fixing will be required.

Painting is measured in m^2 to each side of windows and giving the pane size classification in the description as SMM M60.2.1–4.1.0. Where panes of more than one size occur then the sizes are averaged (SMM M60.M6). External painting must be kept separate and so described (SMM M60.D1). The work is deemed to include the edges of opening lights and portions uncovered by sliding sashes in double hung casements, additional paintwork to the surrounding frame caused by opening lights, cutting in next glass and work on glazing beads, butts and fastenings attached thereto (SMM M60.C4). Work to associated linings and sills is measured as general surfaces (SMM M60.M7).

ADJUSTMENT OF WINDOW OPENINGS

The adjustment of the superficial area of walling, including facework, and internal finishings for the space occupied by the window and frame is followed by the various items, mostly linear, that have to be taken around the opening. The worked examples that follow indicate the method to be followed with a variety of different forms of construction. This work needs to be taken off very carefully and systematically.

WINDOW SCHEDULES

In most cases, when measuring a complete building, it is desirable to compile a window schedule, if this has not already been prepared by the architect, on which all the essential details relating to the windows and finishings generally will be entered. Column headings will cover such matters as location of window, type of window, size of opening (dimensional diagram), ironmongery, glass size classification, type of glass, wall type and thickness, internal finishing, and possibly details at heads and jambs of openings and sills.

In this way all the windows with similar features can readily be seen and measured together. This simplifies the task of taking off and reduces the number of items involved. Care must be taken to enter the correct timesing figures in each case.

WORKED EXAMPLES

Worked examples follow giving the dimensions of a three light wood casement window, a wood bullseye window and metal casements, together with the adjustment of openings in all cases.

WOOD CASEMENT WINDOW

Drawing No. 12

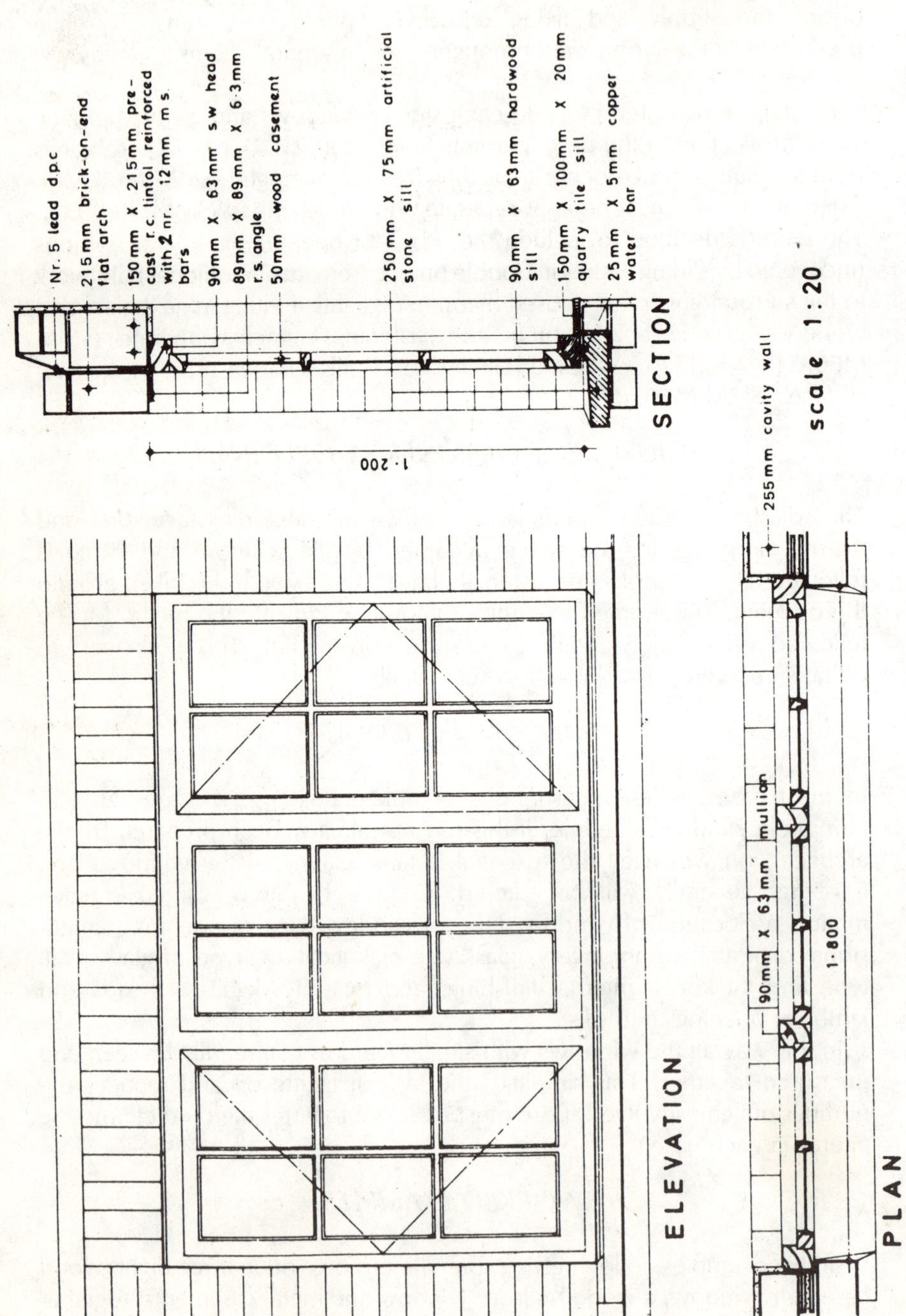

WOOD CASEMENT WINDOW

			wdw. width 1·800 add retns. 2/30 60 1·860	
	1		Wdw. & wdw. fr. as dimnsd. diagrm., wrot. swd., consistg. of 3nr. 50th. reb. & chfd. casts., w. spld. reb. bott. rls. & thro. stiles (2nr. side hg. w. butts (m/s)) & ea. div. into 6nr. panes w. reb. & chfd. glazg. bars, all complyg. w. BS 644 Part 1 & inc. 90 x 63 reb, chfd. & grvd. hd. & jbs. fxd. w. g.w.i. bent cramps 30 x 5 x 200 lg. to bwk. & 2nr. 90 x 63 2ce reb. & 2ce chfd. mulls. & 90 x 63 hwd. sk., 2ce reb. & grvd. sill.	
			[diagram: 1·800 wide x 1·200 high] 2/1·860 3·720 2/1·200 2.400 6·120	
	1.86		Cop. water bar 5 x 25 in sills bedded in mastic.	
	6·12		Bed wd. fr. in c.m. (1:3) & pt. o.s. in mastic.	
2/	1		Pr. 75 p.s. butts & fxg. to swd.	
			Provide the P.C. Sum of £14 for the supply of 2nr. cast. fasteners & stays. & Add for profit.	
			17.1	

EXAMPLE XVII

Wood casements and sash windows together with their frames are enumerated and described, together with a dimensioned diagram (SMM L10. 1.0.1.0). It was not felt that a small dimensioned diagram could adequately cover all the appropriate details in this case and so a full description has been incorporated. Where the ironmongery is supplied with the component, it is included in the description of the window, as are also the method, of fixing (where not at the contractor's discretion).

All timber sizes are deemed to be basic (nominal) sizes, unless stated as finished sizes (SMM L10. D1).
An alternative approach to the measurement of the window would be to enlarge the dimensioned diagram, to incorporate in full the necessary dimensions and descriptions and so reduce the length of the description, which could be considered rather lengthy.
Another approach would be to use a standard window and give the manufacturer's catalogue reference (See SMM L10. M1).

The copper water bar is taken as a linear item with a dimensioned overall cross-section description as SMM P20. 2.1.0.0.
The ironmongery is measured separately as it is not supplied with the component (SMM L10. C2c).
The size and type of butt is given and the nature of the base to which it is fixed, in accordance with SMM P21.1.1.

Prime cost sums may be used to cover the supply of casement fasteners and stays, with provision for the addition of main contractor's profit as SMM A52.1.1–2.1.0.

WOOD CASEMENT WINDOW (Contd.)

2/	1	Fix only cast. fast. in alum. alloy to swd. & Fix only cast. stay in alum. alloy to swd.
		Glazg.
		width: 1.860; less jbs. 2/50 100, mulls. 2/38 76, stiles 3/2/38 228 = 404; 6)1.456 = 243
		ht.: 1.200; less hd. 50, sill 30, top rl. 38, bott rl. 63 = 181; 3)1.019 = 340
3/6/	0.24 0.34	Glazg. w. stand. plain glass, in panes area ≤ 0.15 m², c.s.g. (O.Q.) 3th. to BS.952 & glazg. to wd. w. l.o. putty & sprigs. (In 18 nr. panes)
		Painting
	1.86 1.20	Paintg. glzd. wdws., wd., panes, area ≤ 0.10 m², k.p.s. & (3).
	1.80 1.20	Ditto., ext.
		1.860; 1.200; 2/3.060 = 6.120
	6.12	Paintg. gen. isoltd. surfs., wd. gth. ≤ 300, appln. on site prior to fixg., primg. only. (back of wdw. fr.

This is followed by fixing only items, which are deemed to include preparing the base to receive ironmongery. Some surveyors may prefer to use combined supply and fix items for these fittings.

Calculation of area of individual panes of glass, taking overall dimensions of casement and deducting width of casement members, less width of rebate to receive glass.

Ordinary glazing quality glass to BS 952 is being used. Full particulars are given as required by SMM L40.S1-5, including the kind, quality and thickness of glass, method of glazing and securing and nature of frame or surround.

The pane size classification is given as SMM L40.1.1.1-2, and the number of panes is stated where the area of each pane is ≤ 0.15 m². Where 50 or more panes are identical these are given separately in m², stating the number and size as SMM L40.1.1.1.-2.1. in view of the resultant cost saving.

Painting to windows is measured as glazed windows, on each side of the window, measured flat as SMM M60.M5. The pane sizes are classified in the areas given in SMM M60.2.1-4.1, and where panes are of more than one size, then the sizes are averaged (SMM M60.M6). The painting on glazed windows is deemed to include edges of opening lights, additional painting to the surrounding frame caused by the opening lights and work on attached glazing beads, butts and fastenings (SMM M60.C4). Where the frames and casements are painted in different colours, this would not constitute multi-coloured work as they are separate surfaces (See SMM M60.D2).

External painting is so described, mainly to allow for the extra cost of applying paint to external surfaces (SMM M60.D1).

WOOD CASEMENT WINDOW (Contd.)

Timesing	Dimensions	Squaring	Description	Notes
			Adjust. of opg. ht. 1·200 add arch & lintel 225 sill 75 1·500	Priming has been included in the description, although most windows are delivered to the site already primed, as pink shop primers are of variable quality and soon become weak or powdery allowing water penetration. This would also be influenced by the specification. Priming of back of window frame classified as general isolated surfaces and measured as a linear item as $\leq$ 300 mm girth and including reference to application of primer on site prior to fixing (SMM M60.1.0.2.4).
	1·86 1·50		Ddt. Bk. wall, thickness 102·5, comms. a.b. & Ddt. Ditto. facewk. o.s., multi-col. fcg. bks. a.b. & Ddt. Form cav. to holl. wall, width: 50, a.b.	It is considered desirable to commence with the deduction of the area of walling and finishings. The areas occupied by the arch, lintel and sill are included in the deduction of brickwork and facework. No adjustments have been made for the brickwork displaced by the ends of the lintel and sill, as they are assumed to be offset by the labour in cutting the brickwork, which is deemed to be included under SMM F10. C1b.
	1·86 1·20		Ddt. Pla. to bk. walls a.b. & Ddt. Paintg. g.s., 2ce emulsn. pt. pla. walls.	The deductions of plaster and decorations are for the net area of the opening.
2/	0·03 1·20		Add Bk. wall, thickness: 102·5, multi-col. fcgs. a.b. (reveals	Brickwork in rebated reveals is measured as half-brick wall. They cannot be regarded as projections as they do not constitute attached piers as defined in SMM F10. D9. The cutting is deemed to be included (SMM F10. C1b).
			Arch	
	1·80		Arch, flat bk-on-end, ht. on face: 225, width of exp. soff. 102, fcgs. a.b. (In 1 nr.)	Faced arches are measured in metres giving the height on face and width of exposed soffit, and the shape of the arch, as SMM F10.6.1. The face width of the arch includes the thickness of a mortar joint.
			stl. ∠ len. 1·800 add beargs. 2/100 200 2·000	
	1		Stl. isoltd. structl. membr., plain, 88·9 x 88·9 x 6·27 mm x 8·48 kg/m, 2·00 m lg., r.s. ∠ to BS4 Pt: 1 Table 14 as bk. arch. suppt.	Measured in tonnes as isolated structural members, plain, and stating its use as SMM G12.5.1.1.0.

17.3

WOOD CASEMENT WINDOW (Contd.)

	2.00		Paintg. structl. met., gen. isoltd. surfs., gth. ≤ 300, ext., prep, prime & (3). Lintel len. 1.860 add beargs. 2/215 430 2.290	Painting is measured as a linear item as not exceeding 300 mm girth, and classified as SMM M60.5.1.2.0. No measured item is necessary for building in the ends of the steel angle (See SMM F10.C1b and c).
	1		Precast conc. lintel, 150 x 215 x 2.29 m lg. (1:2:4/20 agg.) reinfd. w. 2 nr. 12 dia. m.s. bars to BS 4449, w. 2 nr. surfs. (300 g/a. gth.) keyed for pla., bedded in g.m. (1:1:6).	Precast concrete lintels are enumerated with a dimensioned description and giving details of materials, mix and reinforcement, and method of bedding (SMM F31.1.1.0.1 and F31.S1-5). The 215 mm bearing at each end allows for a mortar joint. Formwork is deemed to be included with the item (SMM F31.C1).
	2.29 0.38		Dpc, width > 225, hor., cav. tray, single layer of nr. 5 milled lead, w 150 laps, inc. beddg. on 2 nr. bk. skins in c.m. (1:3) to the profile shown on Dwg. 12.	Damp-proof courses forming cavity trays in hollow walls are measured in m^2 and so described as SMM F30.2.2.3.1. No allowance is made for laps (SMM F30.M2).
			Reveals	Labour in returns is deemed to be included in the brickwork rates (SMM F10.C1f), and the facing bricks have already been taken in the walling, hence no additional item is required for faced brick returns to window openings. However, reveals are measured in metres in accordance with SMM F10.11.1.1 as an extra over item where special bricks are used.
2/	1.20		Closg. cavs., width: 50, slates 100 wide, set in c.m (1:3), vert.	Closing cavities is measured in metres, giving the width of cavity, method of closing and plane (SMM F10.12.1.1.0).
2/	1.20		Pla. to walls, width ≤ 300, in 2 cts. a.b., bwk., trowld. fin. (reveals	Plasterwork not exceeding 300 mm wide is so described and measured in metres (SMM M20.1.2.1.0). Plasterwork to sides and soffits of openings is included with the abutting walls (SMM M20.D5), keeping similar work on different types of base separate (SMM M20.S5).
	1.86		Ditto., conc., do. (soff.	

17.4

WOOD CASEMENT WINDOW (Contd.)

Dimensions	Description	Notes
	Rdd. ∠ 1·860 2/1·200 2·400 4·260	
4·26	Rdd. ∠ to pla., rad. 10 – 100.	Rounded angles to plaster are measured in metres as SMM M20. 16 where in the radius range 10 – 100 mm, but where less than 10 mm radius they are not measurable. Plasterwork is deemed to include fair joints to window frames and the like (SMM M20.C1a).
4·26 0·09	Paintg., gen. surfs., seal & 2ce emulsn. pt. pla walls.	The decorations to plastered surrounds to the window are added to the paintwork on the adjoining walls and measured in m², as they do not form isolated areas.
	Sill len. 1·800 add ends 2/100 200 2·000	
2·00	Art. st. sill 250 x 75 sk. wethd. 175 gth. thro. 20 gth. & grvd. 30 gth., w. 2 nr. stoolgs., > 1·50 m lg., bldg. against other wk., & b&p. in g.m. (1:1:6).	A description of the material is required and the sill is measured in metres, giving a dimensioned description as SMM F22.7.1. The description is to include lengths > 1.50 m, building against other work and the number of stoolings. A fuller description of the artificial stone and its mix and colour is likely to be included in a preamble clause or a project specification, where available.
1·86 0·08	Add Bk. wall, thickness: 102·5, comms. in g.m. (1:1:6).	Brickwork measured at back of sill. No rough cutting is measured even although it is not a multiple of half-a-brick, as all rough and fair cutting is deemed to be included in the brickwork rates (SMM F10.C1b).
2·23 0·28	Dpc, width > 225, hor. & vert., single layer of nr. 5 milled lead, inc. beddg. in c.m. (1:3).	The damp-proof course under and around the back of the sill is measured in m² & it is > 225 mm in girth (SMM F30.2.2.1 & 3). The material description is to include the particulars listed in SMM F30. S4 – 6.

17.5

WOOD CASEMENT WINDOW (Contd.)

1.86	Sill red quarry tiles, 20th. b.& p. in c.m. (1:3) w. flush jts. on screeded bed (m/s).	The sill is measured in metres, stating the width as SMM M40.7.0. 1.0 and giving the particulars listed in SMM M40. S1-7. Cutting at ends is deemed to be included (SMM M40. C1d).
1.86	Screed to margin, width : 100, 10 ct.& sd. (1:2) in 1 ct. to rec. q. tiles laid lev. on bwk.	The bed is measured in metres and classified as SMM M10.11.0. 1.0 (margins), as the nearest appropriate item and giving the necessary particulars listed in SMM M10. S1-6.

BULLSEYE	WINDOW		EXAMPLE XVIII
	1	Wdw. fr. as dimnsd. diagrm., 75 x 50 reb. rdd. & grvd. wrot swd. to bullseye wdw., 300 o/a. rad., inc. headg. jts. w. keys & wedges, & fxg. w. g.w.i. bent cramps 30 x 5 x 200 lg. to bwk.	The bullseye window frame is an enumerated item as SMM L10.1.0.1, giving a dimensioned diagram and any other necessary particulars as SMM L10.S1–10. If there had been a bullseye casement inside the frame, this would have been combined with the frame in a single item.
		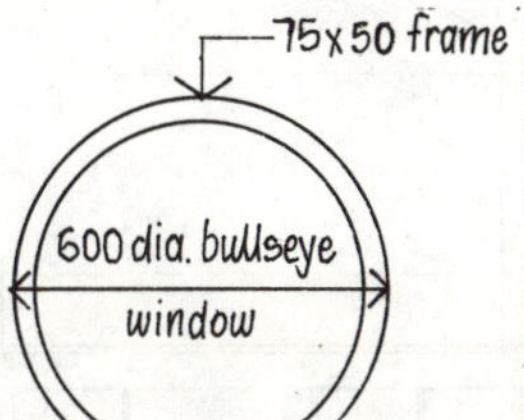	
22/7	0·60	Bed wd. fr. in c.m. (1:3) & pt. o.s. in mastic. 600 less fr. 2/40 80 520	As SMM L10.10.
	0·52 0·52	Glazg. w. stand. plain glass, in irreg. shaped panes, area 0·15–4·00 m², c.s.g. (OQ) 3th. to BS 952 to wd. w. l.o. putty & sprgs.	The glass is measured the smallest rectangular area from which the irregular shape can be obtained (SMM L40.M3), and measured in accordance with SMM L40.1.1.2.2, giving the appropriate area of pane range. Curved cutting is deemed to be included in the glazing item (SMM L40.C1).
22/7	0·30 0·30	Paintg. glazd. wdws., wd., panes area: 0·10 – 0·50 m², k.p.s. & ③ & Ditto., ext.	Painting to casement is measured each side of the window in m² over frames and glass and stating the area of pane classification as SMM M60.2.2.1.0. The circular window is not considered to come within the classification of irregular surfaces. External painting is so described (SMM M60.D1).
		Adjust. of opg.	
22/7	0·30 0·30	Ddt. Bk. wall, thickness: 102·5, comms. in g.m. (1:1:6). & Ddt. Ditto. in facewk. o.s., in red fcg. bks. a.b.	Deduction of hollow wall and finishings for area occupied by bullseye window. Note the use of the letters 'a.b.' to refer back to earlier full descriptions of the items concerned.
		18.1	

BULLSEYE WINDOW Drawing No. 13

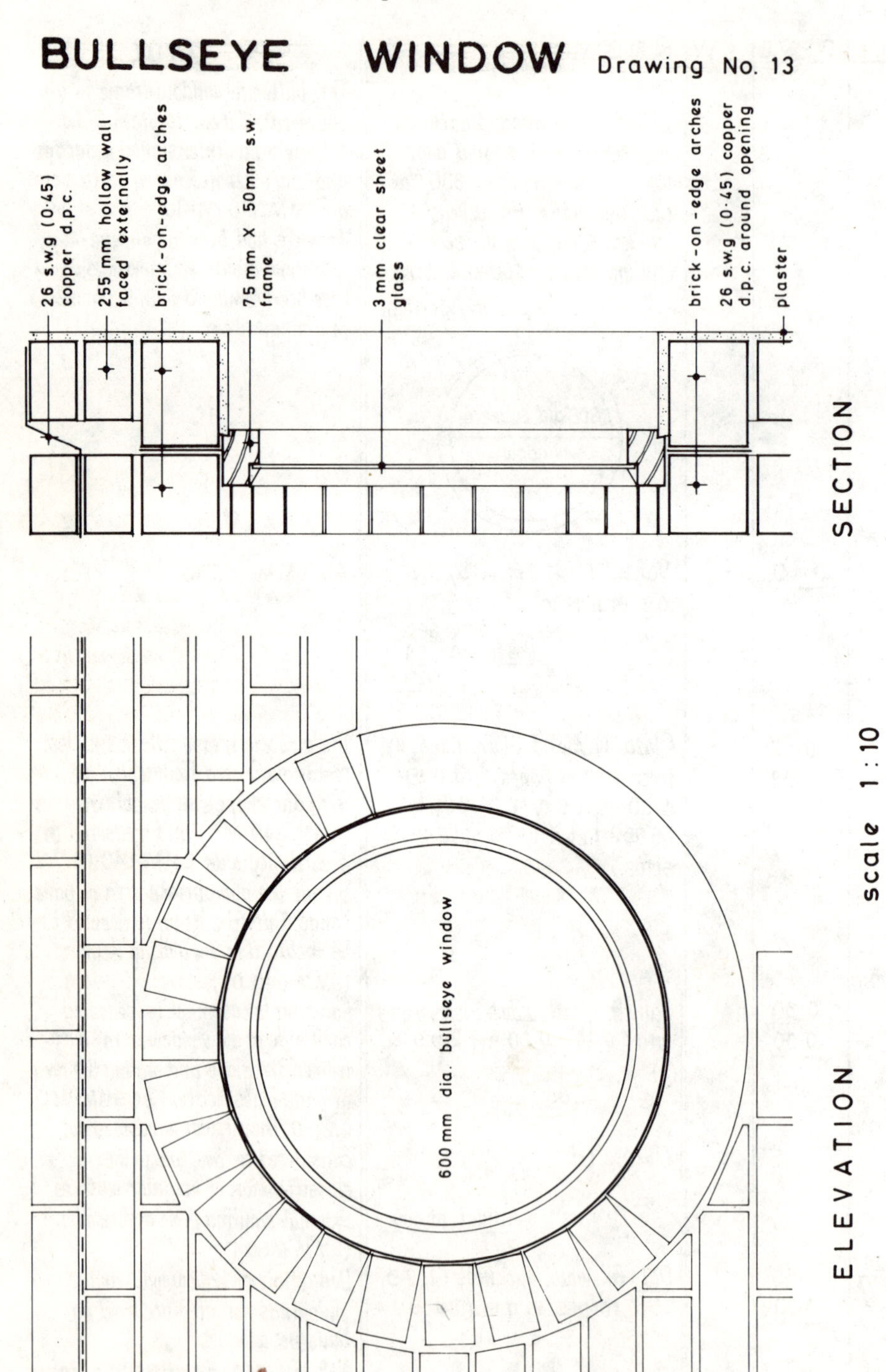

BULLSEYE WINDOW (Contd.)

22/7	0.30 0.30		Ddt. Form cav. to holl. wall, width : 50 a.b. & Ddt. Pla. to bk. walls a.b. & Ddt. Paintg. g.s., 2ce emulsn. pt. pla. walls.	Continuation of adjustments for area occupied by bullseye window.
			Arches 600 add 2½/112 112 mean diam. 712 2)712 mean rad. 356	Note use of subheading for arches.
22/7	0.71		Arch, circ., to 356 mean rad. ht. on face : 112, width on exp. soff. 70, fcgs. a.b. (In 1 nr.).	The arch length is taken as the mean girth on face, as SMM F10.M6, and the face width includes a mortar joint. Faced arches are measured in metres, giving the height on face and width of the exposed soffit, and the shape of the arch (SMM F10.6.1). Curved work is so described with the radii stated (SMM F10.M4).
22/7	0.71		Arch, circ. to 356 mean rad. bk-on-edge, ht. on face : 112, width on exp. soff. 150, comms. a.b. (In 1 nr.).	Arch in common bricks is measured in a similar manner to the faced arch in accordance with SMM F10.6.1, and no separate item is required for closing the cavity.
22/7	0.71 0.11		Ddt. Bk. wall thickness : 102.5, comms. in g.m. (1:1:6) a.b. & Ddt. Ditto. facewk. o.s. in red fcgs. a.b. & Ddt. Form cav. to holl. wall, width : 50, a.b.	Fair cutting is deemed to be included in the brickwork rates (SMM F10.C1b). Centering to arches is also deemed to be included (SMM F10.C1g). Deduction of the brickwork and cavity for the area occupied by the two arches.

BULLSEYE WINDOW (Contd.)

Timesing	Dimensions	Squaring	Description	Explanation
22/7/	0.71 0.11		Dpc, width ≤ 225, vert., curved, single layer of 26 gauge cop. bedded in c.m. (1:3), w. 100 laps.	Damp-proof course is measured in m², classified as ≤ 225 mm wide as SMM F30.2.1.1.0, and giving the particulars listed in SMM F30. S4-6. Curved work is so described (SMM F30. M1) but it is unnecessary to state the radius.
22/7/	0.60		Pla. to walls, width ≤ 300, curved to 300 rad., in 2 cts. a.b., bwk. trowld. fin. (reveals	Curved plasterwork is separately described, giving the radius (SMM M20.M5), and as the width does not exceed 300 mm, this is stated in the description of the item and it becomes a linear item (SMM M20.1.2.1.0).
22/7/	0.60		Rdd. ∠ to pla., curved to 300 rad., rad. 10 – 100.	Rounded angles to plaster are measured in metres if in the 10 – 100 mm radius range. Curved work is given separately stating the radius (SMM M20. M5). No item is required for finishing the plaster against the window as fair joints are deemed to be included (SMM M20.C1a).
22/7/	0.60 0.15		Paintg. gen. surfs., prep. & 2ce emulsn. pt. pla. walls.	This is taken as a superficial item to be added to the walls with the same finish. It is unnecessary to keep curved painting work separate from straight work, as there is no additional labour involved.
			width inner skin 102 dia. across cav. 100 outer skin. 102 304	
	1.02 0.30		Dpc, width > 225, curved, stepped, single layer of 26 gauge cop. inc. beddg. on 2 nr. bk. skins in g.m. (1:1:6) and to profile shown on Dwg. 13.	Combined dpc and cavity gutter measured in m² and exceeds 225 mm wide, and giving the particulars listed in SMM F30.2.2.4.0. Curved work is so described (SMM F30. M1), and it seems beneficial to refer the estimator to the drawing, in order that he shall appreciate the shape of the component.

METAL CASEMENT WINDOWS IN STONE SURROUND EXAMPLE XIX

		<u>Wdws.</u> Provide the PC Sum of £90 for the supply of 3 nr. met. casts., inc. 1 nr. w. semi-circ. hd. & opg. lt. & <u>Add</u> for profit	Metal casements are covered by enumerated items, accompanied by a dimensioned diagram as SMM L11.1.0.1. In this example, a prime cost sum has been inserted in accordance with SMM A52.1.1–2.1, when it is necessary to give a description of the item to be supplied and to make provision for the addition of profit by the main contractor.
2/	1	Fix only met. cast. & fr. 325 x 925 in size into st. surrd. inc. scrg. to lead plugs & beddg. fr. & ptg. b.s. in mastic.	Fixing only metal casements is enumerated, including screwing to lead plugs and bedding and pointing the frame, following the supply item.
	1	Ditto. w. semi-circ. hd., 625 x 1225 in size o/a, inc. do.	Mortices in stonework and metal cramps, slate dowels, metal dowels, lead plugs and the like are deemed to be included in the stonework rates (SMM F21.C1c&e), and their type and positioning will normally be included in a preamble clause or the project specification, where available.
2/	0·28 0·89 0·56 0·83	Glazg. w. stand. plain glass, panes, areas: 0·15–4·00 m², float glass, 6th. to BS 952 to met., w. met. putty & pegs.	Full particulars of the glass and glazing are given in accordance with SMM L40. S1–5, and pane size classification as SMM L40.1.1.2.0. The dimensions of the glass have been scaled in this example, as a large scale drawing has been supplied.
	0·57 0·29	Ditto. in irreg. shaped panes, area: 0·15–4·00 m² in do.	Irregular panes are measured the smallest rectangular area from which they can be obtained and are so described (SMM L40.M3.). Curved cutting to the glass around the semi-circular head is deemed to be included (SMM L40.C1), and does not therefore require separate measurement.
	0·63 0·93	<u>Paintg.</u> Paintg. glazd. wdws., met., panes area: 0·50–1·00 m², prep., ① cal. plumb. primer & ③. & Ditto. ext.	Painting to metal casements is measured over frames and glass in m² giving the appropriate pane area classification as SMM M60.2.2/3.1.0, and the relevant particulars listed in SMM M60. S1–8.
2/ ½/22/7/	0·33 0·93 0·31 0·31	Ditto. panes, area: 0·10–0·50 m², do. & Ditto. ext.	The external painting dimensions are slightly larger, but as the difference is so small it has been ignored. External painting is kept separate and so described (SMM M60.D1).

19.1

METAL CASEMENTS IN STONE SURROUND

Drawing No. 14

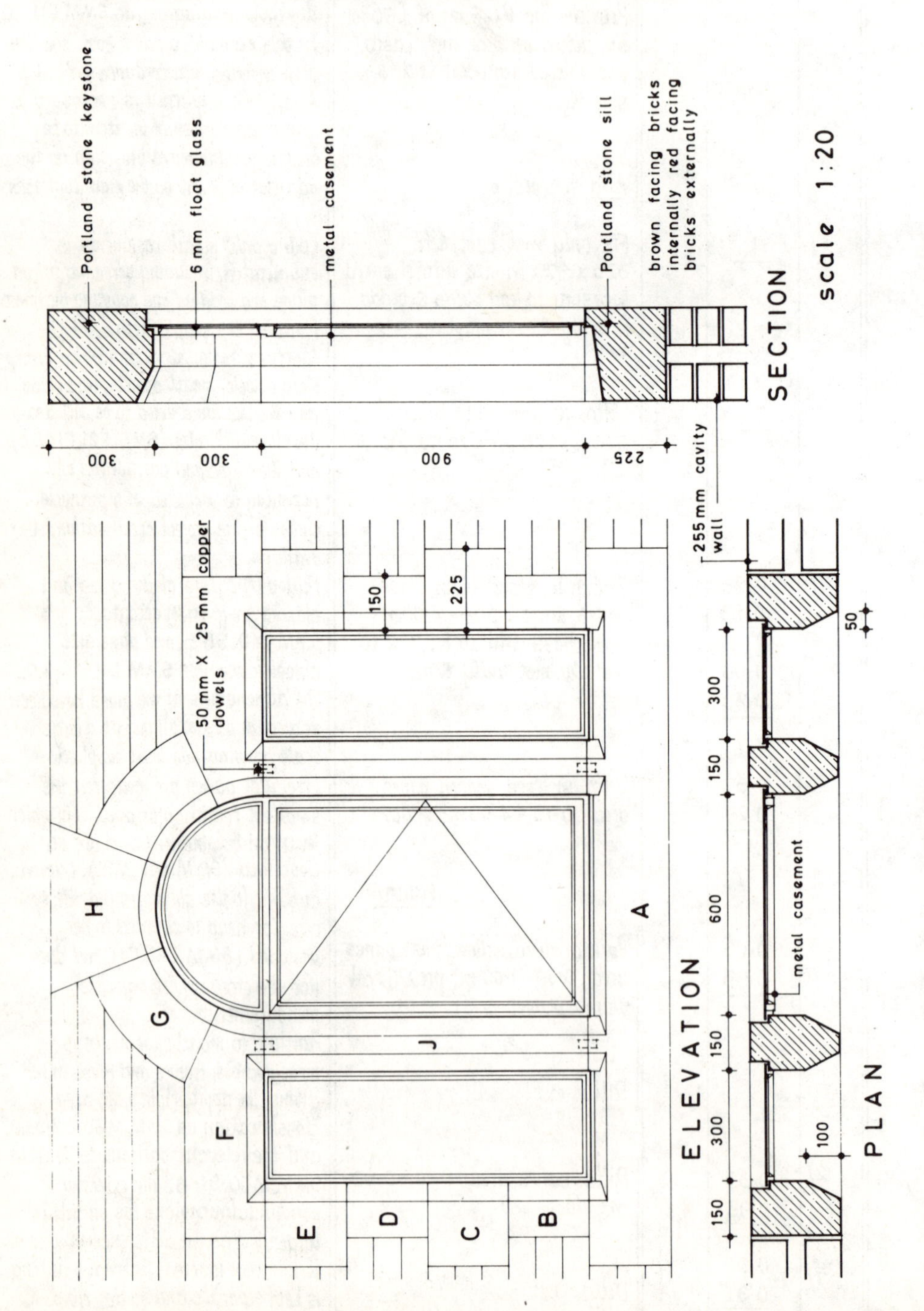

METAL CASEMENT WINDOWS IN STONE SURROUND (Contd.)

			Adjust of opg.	Painting of edges of opening lights are deemed to be included in the measurement of the windows (SMM M60. C4a).
			P. st. (Whitbed), rubd. on exp. faces, b. & j. in st. dust mo. consistg. of 2pts. wh. ct., 5pts blue lias lime & 7pts st. dust, ptg. w. a neat flush jt. & cleang. down on completn. in st. wdw. dressgs. to bk. faced walls.	The description of the stone and jointing is given in a preamble clause to avoid lengthy billed descriptions, with the appropriate particulars as SMM F21.S1–10, which will also include details of the type and positioning of metal cramps, slate dowels, metal dowels, lead plugs and the like, all of which are deemed to be included in the stonework rates (SMM F21.C1e). External elevations will also show the materials used (SMM F21.P1b).
			len. of sill 2/225 450 2/300 600 2/150 300 600 1.950	Stone dressings are those in walls of other materials, as in this example (SMM F21.D2), and this is incorporated in the heading to avoid inclusion in each item of stonework. It is good practice to number or letter the stones for ease of reference, as given on the drawing. Alternatively, dimensioned diagrams can be incorporated in the Bill.
	1.95		Sill, 255 x 225, sk. wethd., faced b.s., w. 4nr. stoolgs., blks. > 1.50 m lg. (A)	The descriptions of stones commence with the function of the stone, then size, and finally the labours to make up full dimensioned descriptions. The sill is measured in metres with a dimensioned description (SMM F21. 7.1.0.5). Stoolings are included in the item description, stating the number. Blocks > 1.50 m long or 0.50 m³ are so described. It is considered that the term 'building against other work' does not apply unless the stone is backed with another material.
2/2/	0.23		Jamb sts., isoltd., 150 wide on face & 255th, reb. & chfd., faced b.s. (B & D) & Ditto. 225 wide on face & 255 th., do. (C & E)	Jamb stones are taken as linear items as SMM F21.11.3.1, giving a dimensioned description and stating whether isolated or attached. Relevant particulars from the fourth column of SMM 7 are also extracted.

19.2

METAL CASEMENT WINDOWS IN STONE SURROUND (Contd.)

2/	0.63		Lintels, (in 2nr.) 255 x 225, reb. & chfd., ea. sk. jtd. circ. sk. circ. chfd. & circ. reb. at 1nr. end & ea. w. 2nr. stoolgs., faced b.s. (F)	Lintels are measured in metres giving a dimensioned description as SMM F21.6.1. It seems advisable to indicate that the linear item comprises two separate components.
				The following definitions of masonry terms may be helpful, although many of them are not included in the labours listed in SMM7 (F21). circular: curved convex surface; circular sunk: curved concave surface; sunk jointed: sunk face adjoining brickwork or masonry; circular jointed: ditto with curved convex surface; circular sunk jointed: ditto with curved concave surface.
	9.60		Arch (1nr.) comprisg. 3nr. sts., hts. on face varying from 225 – 300, width of soff. 225, semi-circ. shape, sk. jtd., circ. sk., circ. reb. & circ. chfd., w. rad. bed, faced b.s. (G & H)	The arch is measured in metres on the mean girth on face (SMM F21.M8), and the description includes the number of arches, height on face, width of soffit and shape of arch (SMM F21. 24.1). The various labours have also been included. Alternatively, the arch could be enumerated. It is assumed that the natural bed will be on the line of the lowest joint (giving one sunk joint only) and the sizes represent the mean dimensions (SMM F21.M1). Stones are deemed to be set on their natural beds unless otherwise described (SMM F21.S9).
2/	0.90		Mulls. 255 x 150, 2ce reb. & 2ce chfd., faced b.s., & fxd. w. cop. dowels. (J)	Mullions are taken as linear items as SMM F21.8.1. The copper dowels and mortices in stonework to receive them are deemed to be included in the stonework items (SMM F21.C1c&e), but the method of jointing should be stated as SMM F21.S7. A description of the dowels can be given in a preamble clause or by reference to a project specification.

19.3

METAL CASEMENT WINDOWS IN STONE SURROUND (Contd.)

Timesing	Dimensions	Squared	Description	Notes
	1		Centrg. to st. arch, semi-circ., 600 span & 255 wide on soff.	Enumerated item as SMM F21. 36.1.1.0.
			Adjust. of Bwk. 1.950 less 2/75 150 1.800	
	1.95 0.23		Ddt. Bk. wall, thickness: 102.5, facewk. o.s. in red fcgs. a.b. (sill	Deduction of brickwork and facework for the area occupied by the windows and stone surround. Each dimension is taken to the nearest 10 mm, with 5 mm and over regarded as 10 mm, in accordance with SMM General Rules 3.2, even although this results in a slight over-measurement of the items deducted.
2/	1.80 0.23		& (B & D	
2/	1.95 0.23		Ddt. Ditto., facewk. o.s. in brown fcgs. a.b. & (C & E	
2/	0.39 0.23		Ddt. Form cav. to holl. wall, width: 50 a.b. (lintels	The void is measured up to the outer edge of the semi-circular stone arch (scaled from drawing).
½/22/7	0.53 0.53		(semi-circ. st. arch.	Note locational notes in waste.
	0.44 0.09		(proj. to key st.	All fair cutting to brickwork is deemed to be included in the brickwork rates (SMM F10. C1b).

11 Measurement of Doors

ORDER OF MEASUREMENT

The measurement of doors can be broadly sub-divided into internal and external doors, and the dimensions of each of these two classes of door broken down into: (1) door and (2) adjustment of opening.

The door dimensions will include any ironmongery, glazing and painting required together with the frame or lining from which the door is hung. When adjusting the opening for an external door care must be taken to cover all the appropriate items by adopting a logical order of taking off such as deduction of wall, external and internal finishings, and measurement of external head or arch, internal head or lintel, external reveals or margins, where necessary, internal reveals and threshold. It is sometimes necessary to adjust the skirting and flooring in the door opening, where these have not been covered with finishings or floors, and to measure steps leading up to the door.

Where sidelights and/or fanlights are to be provided adjacent to the door, these will be measured at the same time as the door.

When taking off the dimensions of a number of doors, it is advisable to prepare a schedule in the first instance, detailing such information as location or number of each door, size and type of door, details of ironmongery, size of frame or lining, thickness and type of wall or partition, finishings to both sides of wall, details of head of opening, reveals and threshold. In this way the task of taking off the doors will be greatly simplified and similar items suitably grouped as the work proceeds, so avoiding duplication of items. The schedule will be marked as each item is extracted and entered on the dimensions paper.

Most surveyors take off the doors first and then follow with the adjustment of the openings; the main justification for this procedure is that the size of the door determines the size of the frame or lining, which in its turn decides the size of the opening.

DOORS

The principal rules governing the measurement of the doors themselves are contained in SMM work section L20. Doors are enumerated with a dimensioned diagram in accordance with SMM L20.1.0.1.0, as illustrated in example XXI. However, it was not considered necessary to provide a dimensioned diagram in the case of the straightforward flush door in example XX. Each leaf of a multi-leafed door is counted as one door (SMM L20.M2). With glazed doors, the method of glazing, such as use of wood beads, and method of securing by brads or screws, and gasket where provided, are included in the description of the glass (SMM L40.S3–4). The glazing beads themselves are included in the description of the door. The door item is deemed to include glazing which is supplied with the component (SMM L20.C3e). Stock pattern doors can be described by reference to a manufacturer's catalogue or the appropriate British Standard, and in the case of a composite item assembled off site it can include the associated frame or lining (SMM L20.M4). Fitting and hanging of doors are deemed to be included with the items (SMM L20.C1). All sizes of timber are nominal sizes unless stated as finished sizes (SMM L20.D1).

The provision and fixing of units or sets of ironmongery is covered by enumerated and fully described items and the nature of the base must be stated (SMM P21.1.1 and P21.S1–3). Where the supply only of the ironmongery is covered by a prime cost sum, separate fixing items containing the necessary descriptive particulars are needed. Alternatively, a basic price approach may be adopted in a single combined provision and fixing item as illustrated in example XX in this chapter. Ironmongery where supplied with the component is deemed to be included (SMM L20.C3e).

Painting to unglazed doors is classified as to general surfaces (SMM M60.1.0.1.0), whereas painting of glazed doors is classified as glazed doors in accordance with SMM M60.4.1–4.1.3. The latter contains the appropriate pane area classification and will often be described as partially glazed as in example XXI, where only the top panel of the door is open and glazed. Painting to unglazed panelled and matchboarded doors normally have their areas timesed by $1\frac{1}{9}$ to allow for the additional area in moulded surfaces and door edges as listed in SMM M60.M2. The multiplier should be increased to $1\frac{2}{9}$ for a door with large bolection mouldings. With flush doors it is better to take the full area of door surfaces, including top and side edges, thus avoiding the need for the use of a multiplier.

DOOR FRAMES AND LINING SETS

Door frames and lining sets are grouped together and are measured in metres, giving a dimensioned overall cross-section description of each member, separating jambs and heads and stating the number of sets (SMM L20.7.1–2.1.0). The length of door frame or lining will comprise the extreme lengths of each section. SMM L20.7.1–2.1.1 requires attention to be drawn to the incidence of repetition of identical items (for example, 26 nr identical sets), so that the advantage of any possible cost savings can be secured. It is necessary to state the number of lengths of sills, mullions and transoms, even although they are measured in metres (SMM L20.7.3–5.1.0).

The method of fixing of door frames and lining sets, where not at the discretion of the contractor, is included in the item descriptions, in accordance with SMM L20.S8. The method of fixing may take a variety of different forms, such as dowels, cramps, fixing bricks or grounds, and a selection of these are contained in the worked examples that follow.

Architraves are measured in a similar manner to skirtings and picture rails in metres, giving a dimensioned overall cross-section description (SMM P20.1.1). Ends, angles, mitres, intersections and the like are deemed to be included except on hardwood items >0.003 m^2 sectional area (SMM P20.C1).

ADJUSTMENT OF DOOR OPENINGS

The golden rule in measuring this work is to adopt a logical sequence, working systematically through the adjustment from the main area of walling and finishings to the edges of the opening from top to bottom, and both inside and out in the case of an external door. This is illustrated in the worked examples in this chapter.

WORKED EXAMPLES

Worked examples are now given which illustrate the method of taking off the dimensions for internal and external doors, including the adjustment of the openings.

INTERNAL DOOR

EXAMPLE XX

Timesing	Dimensions	Squaring	Description
			<u>Door</u>
	<u>1</u>		Dr. 726 x 2040 x 40th., flush semi-solid cored faced b.s.w. plywood 5 th., glued to the core under pressure & lipped & edged w. hwd. strips to both vert. edges to BS 459 Pt. 2A.
			726 2/2.040 4.080 4.806
2/	0.73 2.04		Paintg. gen. surfs., wd., k.p.s. & (3).
	4.81 0.04		(edges
	<u>1</u>		Pr. of 100 p.s. butts & fxg. to swd.
			&
			Mors. latch & lever furn. in alum. alloy <u>basic price £11 the set</u> & fxg. to swd.
			<u>Dr. ling. set (1 nr.)</u> 2.040 <u>add</u> tongue 13 2.053
			726 <u>add</u> passgs. 2/25 50 776
2/	2.05		Jamb 245 x 38 wrot swd. 2ce reb., tgd. at ∠s., fxd. w. 50 x 25 sn. swd. frd. grds. plugd. to masonry.
	0.78		Head 245 x 38, ditto.
			<u>Archves.</u> 2/2.065 4.130 <u>add</u> 776 angles 2/2/87.5 350 5.256 <u>less</u> plinths 2/125 250 5.006
			20.1

The flush door has been enumerated and described. The dimensioned diagram prescribed by SMM L20.1.0.1.0 is so simple that it has been omitted, as the leading dimensions can so easily be given in the written description. Fitting and hanging the door is deemed to be included in this item (SMM L20. C1). In the case of a standard door it would be sufficient to make reference to the manufacturer's catalogue (SMM L20. M1).

Painting on both faces and top and side edges of the door are measured in m² and classified as painting to general surfaces as SMM M60.1.0.1.0. Work is deemed to be internal unless otherwise described (SMM M60. D1).

Units or sets of ironmongery are each enumerated separately. The provision and fixing can be a single combined item but in this case the basic price approach has been adopted in place of a prime cost sum. The nature of the base to which the ironmongery is fixed shall be included (SMM P.21.1.1).

A general preferred approach would probably be to insert a single prime cost for the supply of all ironmongery with separate enumerated fixing items. The number of door lining sets is to be stated (SMM L20.7).

Jambs and head of door lining sets are each provided with a dimensioned overall cross-section description and given separately in metres as SMM L20.7.1-2.1.0.

All timber sizes are deemed to be nominal unless stated as finished sizes (SMM L20. D1).

The method of fixing the linings is included in the item descriptions, where not at the contractor's discretion (SMM L20. S8).

INTERNAL DOOR

Drawing No. 15

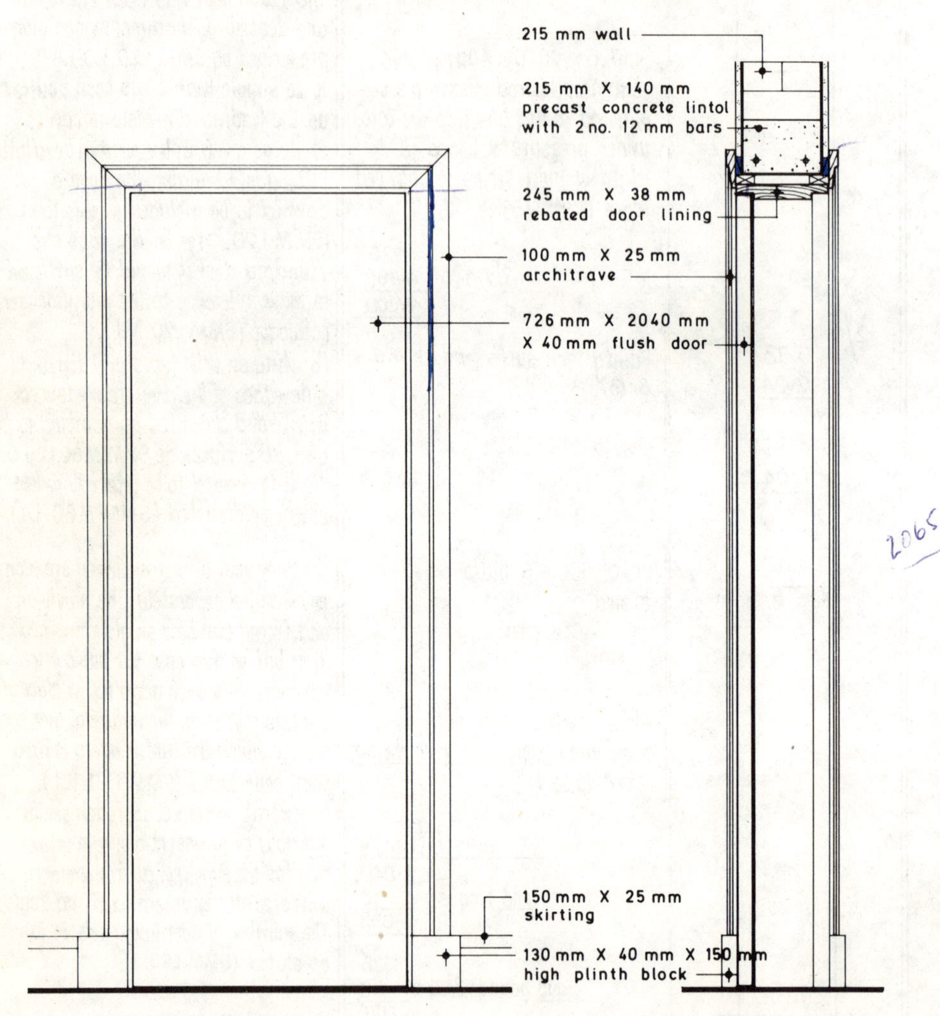

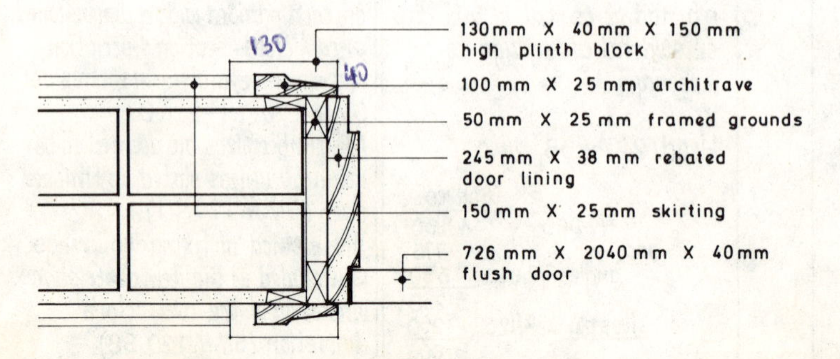

PLAN scale 1:10

INTERNAL DOOR (Contd.)

				Plinth blocks are deducted from the length of the architrave as they are measured separately (depth of plinth block less 25 mm housing for architrave).
2/	5.01		Archve. 100 x 25 wrot, swd. 2ce spld. & reb., fixd. w. 50 x 12 sn. swd. spld. grd. plugd. to bwk.	Linear item to both sides of door opening as SMM P20.1.1.0.0, giving a dimensioned overall cross-section description. The method of fixing is included as SMM P20.S8. Ends and mitres to architraves are deemed to be included, as the architrave is not in hardwood > 0.003 m^2 sectional area (SMM P20.C1).
2/	2		Pl. blk., wrot swd., 130 x 40 x 150 hi., morticed to rec. end of 100 x 25 archve.	Plinth blocks are enumerated with a dimensioned description (SMM P20.9.1.0.0).
			Paintg. av. len. 4.130 776 4.906 less pl. blks. 2/150 300 4.606 width ling. 245 add edges 2/25 50 archves. 2/150 300 595 add pl. blks. 2/55 (at base) 110 705	Build up of the combined area of lining and architraves for painting purposes. As the combined width exceeds 300 mm, this becomes a superficial item (SMM M60.1.0.1.0) classified as to general surfaces.
	4.61 0.60		Paintg. gen. surfs., wd., K.p.s. & (3). (lings. & archves.	
2/	0.71 0.15		(lings. & pl. blks.	
2/	5.01 4.91		Paintg. gen. isoltd. surfs., wd., gth. ≤ 300, appln. on site prior to fixg., primg. only. (archve. (ling.	Linear painting items to back of members as ≤ 300 mm girth, and description includes application on site prior to fixing as SMM M60.1.0.2.4.

20.2

INTERNAL DOOR (Contd.)

Timesing	Dimensions	Description	Side notes
		Adjust. of opg. width 726 add lings. (less reb.) & grds. 2/50 100 826 ht. 2.040 add ling. & grds. 50 2.090 add conc. lintel 150 2.240	The area to be deducted includes the thickness of linings (less rebate for door) and grounds. The depth of the concrete lintel is included in the brickwork deduction, as no adjustment need be made for the ends of the lintel, as they are assumed to be offset by the labour in cutting the brickwork, which is deemed to be included under SMM F10.C1b.
	0.83 2.24	Ddt. Bk. wall, thickness: 102.5, comms. in g.m. (1:1:6).	It is considered unnecessary to deduct the decorations for the small area occupied by the architrave and skirting beyond the limits of the opening.
2/	0.83 150 – 2.09	Ddt. Pla. to bk. walls a.b. & Ddt. Paintg. gen. surfs., prep. & 2ce emulsn. pt. pla. walls.	
		Lintel 826 add beargs. 2/100 200 1.026	Precast concrete lintels are enumerated with a dimensioned description as SMM F31.1.1.0.1, and giving materials, mix and reinforcement details.
	1	Precast conc. lintel 140 x 215 x 1.026 lg. (1:2:4/20 agg.) reinfd. w. 2 nr. 12 dia. m.s. bars to BS 4449, w. 2 nr. surfs. (300 a/a gth.) keyed for pla. & bedded in g.m. (1:1:6).	Moulds are deemed to be included with the item (SMM F31.C1). As an alternative the concrete in the lintel could be described by strength rather than by mix, such as 21.00 N/mm².

If skirtings are adjusted with the doors, the dimensions would be as follows:

dr.	726
add pl. blks. 2/130	260
	986

Timesing	Dimensions	Description	Side notes
2/	0.99	Ddt. Sktg. 25 x 150 wrot swd. mo., fxd. w. grds. a.b. & Ddt. Paintg. gen. isoltd. surfs., wd., gth. ≤ 300, k.p.s. & ③.	The skirting is measured in metres, giving a dimensioned overall cross-section description as SMM P20.1.1.0.0. Fitted ends to skirtings are deemed to be included and do not require separate measurement (SMM P20.C1).

EXTERNAL	DOOR			EXAMPLE XXI
			Door	Work systematically through the dimensions for the door, glazing, ironmongery, frame and painting, and then cover all aspects of the adjustments for the opening.
	1		Dr. as dimnsd. diagrm. wrot swd., 2nr. pans of ext. qual. plywd., chfd. o.s. & plant mldd. o.s. w. 1nr. pan reb. for glass (m/s), inc. 15x10 swd glazg. bds, bott. rl. holl. reb. for water bar (m/s), & 2ce spld. & thro. weather bd. tgd. to bott. rl.	The door is enumerated with a dimensioned diagram, which enables the description to be confined mainly to kind and quality of materials and ancillary labours (SMM L20.1.0.1.0). Glazing beads are included in the description of the door.

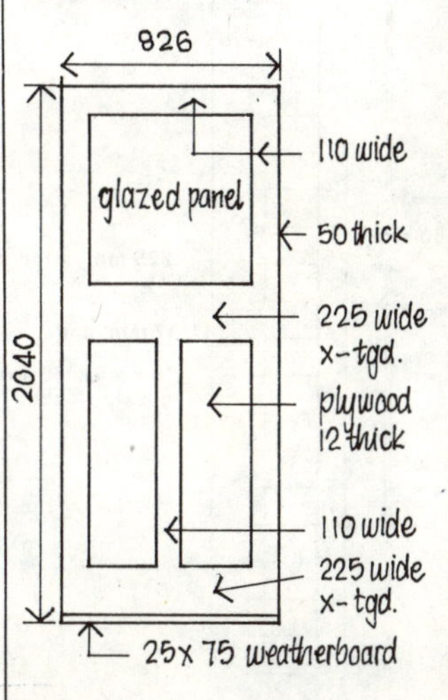

	0·83		5 x 40 x 826 lg. g.m.s. water bar, bedded in c.m. (1:3) in prepd. grve. in art. st.	The water bar is taken as a linear item, with a dimensioned overall cross-section description as SMM P20.2.1.0.0. The groove in the artificial stone threshold will be included in the description of the threshold which is measured later.
			Glazg.	
	0·63 0·63		Glazg. w. stand. plain glass, panes, area: 0·15 – 4·00 m², fluted sheet glass 4th. to BS952 to wd. w. swd. glazg. beads (m/s) secured w. brads & mastic.	Full particulars of the glass and glazing and the pane size classification are given as SMM L40.1.1.2.0 and L40.S1–5.
			Ironmongery	
	1½		Prs. of 100 p.s. butts & fxg. to swd.	Units or sets of ironmongery are enumerated separately as fixing items, with the exception of the butts.

21.1

EXTERNAL DOOR

Drawing No. 16

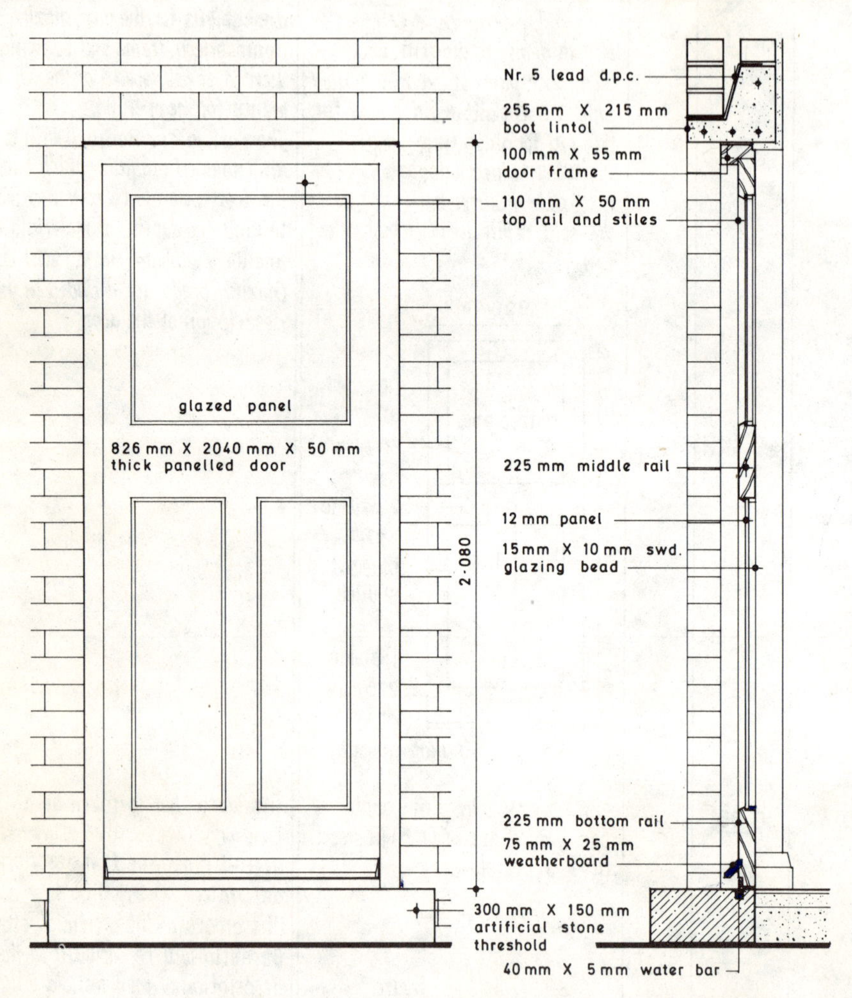

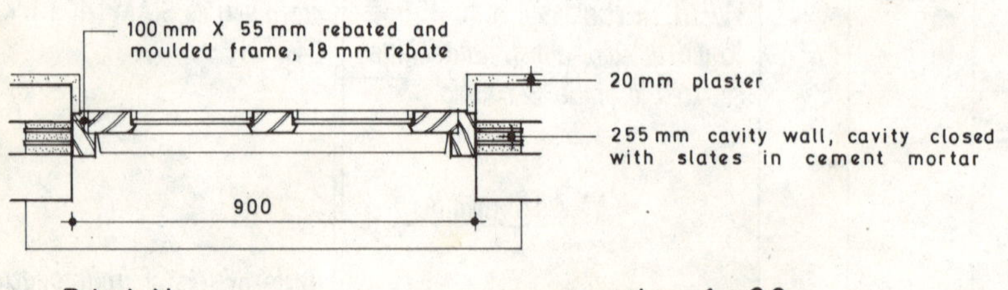

scale 1 : 20

EXTERNAL DOOR (Contd.)

Timesing	Dimension	Description	Notes
	1	Fix only mors. lock & lever furn. in alum. alloy to swd. & Fix only letter plate & knocker in alum. alloy to BS 2911 to swd., inc. formg. aperture. & Fix only 150 lg. brass barrel bolt & soc. to swd.	There will be a comprehensive prime cost item for the supply of all ironmongery to items of joinery.
		Dr. fr. set (1 Nr.)	Door frame sets are grouped together stating the number of sets (SMM L20.7).
2/	2.08	Jamb 100 x 55 wrot swd. 2ce reb. & mo., fxd. w. g.m.s. bent cramps, 30 x 3 x 250 gth. & m.s. dowels, 15ϕ x 100 lg. 900 add horns 2/75 150 1.050	Jambs and head are each measured separately in metres, giving dimensioned overall cross-section descriptions as SMM L20.7.1-2.1.0, and including the method of fixing where not at the contractor's discretion (SMM L20.S8). Also note the requirement to give the number of identical items to indicate the extent of repetitive work; this particularly applies to standard frames and lining sets.
	1.05	Head 100 x 55, wrot swd., 2ce reb. & mo. 2/2.080 4.160 900 5.060	The mortices in the artificial stone for the dowels are deemed to be included in the stonework item (SMM F22.C1c).
	5.06	Bed wd. fr. in c.m. (1:3) & pt. o.s. in mastic. & Paintg. gen. isolated surfs., wd., gth. ≤ 300, appln. on site prior to fxg., primg. only.	Linear item of bedding and pointing frame as SMM L20.10. Priming the back of the frame is classified as SMM M60.1.0.2.4.
		Painting	
	0.83 2.04	Paintg. glazed drs., wd., panes, area: 0.10 - 0.50 m², partially glazed, k.p.s. & (3). & Ditto., ext.	Painting on wood glazed doors is measured in m², on each side, including edges of door, measured flat, stating the pane area classification and including reference to partial glazing as SMM M60.4.2.1.3.

21.2

EXTERNAL DOOR (Contd.)

Dimensions	Description	Notes
5.06	Paintg. gen. isoltd. surfs., wd., gth. ≤ 300, k.p.s. & (3). & Ditto., ext. (fr.	Painting of wood frames is measured in metres as general isolated surfaces as ≤ 300 mm girth, as SMM M60.1.0.2.0. External painting is kept separate and so described (SMM M60. D1).
	Adjust. of Opg.	
0.90 2.08	Ddt. Bk. wall thickness : 102.5, comms. in g.m. a.b. & Ddt. Ditto. facewk. o.s. in g.m. a.b. & Ddt. Form cav. to holl. wall, width : 50, a.b. & Ddt. Pla. to bk. walls a.b. & Ddt. Paintg. g.s. prep. & 2ce emulsn. pt. pla. walls a.b.	Deduction of brickwork and finishings for area occupied by door and frame, adopting a logical sequence of items.
	Lintel 900 add beargs. 2/100 200 1.100	Build-up of length of lintel in waste. The order of dimensions in descriptions shall generally be that of length, width and height (SMM General Rules 4.1). In this example the sequence given is considered more appropriate, with the length specifically indentified.
1	Precast conc. lintel 255 x 215 x 1.100 lg. boot type (1: 1½ : 3/20 agg.) reinfd. w. 4nr. 12 dia. m.s. bars & 6 nr. 6 dia. stirrups to BS 4449, w. 2nr. surfs. (300 o/a gth.) keyed for pla. & 2 nr. surfs. (190 o/a gth.) fin. smth., & 2nr. reb. ends.	The lintel forms an enumerated item with a dimensioned description including details of materials, mix and reinforcement as SMM F31.1.1.0.1 and F31. S1–5. Moulds are deemed to be included in the item (SMM F31.C1).
	d.p.c. len. / width 1.100 outer skin 102 add ends 2/75 150 across cav. 165 1.250 inner skin 102 369	
1.25 0.37	Dpc, width > 225, stepped, cav. tray, single layer of Nr. 5 milled lead w. 150 laps inc. beddg. on 2 nr. bk. skins in c.m. (1:3).	The damp-proof course will need to overhang the lintel at each end. Cavity gutter or tray is measured in m², giving the details prescribed in SMM F30.2.2.4.1.

21.3

EXTERNAL DOOR (Contd.)

Timesing	Dimensions	Description	Notes
	0.90 0.23	Ddt. Bk. wall, thickness: 102.5, comms. in g.m. a.b.	Deduction of brickwork and cavity for the boot lintel, but making no allowance for the ends, where the brickwork displaced is regarded as offset by the additional labour.
	0.90 0.15	Ddt. Form cav. to holl. wall, width: 50, a.b.	
	0.90 0.08	Ddt. Bk. wall, thickness: 102.5, facewk. o.s. in g.m. a.b.	No additional item is required for facework to reveals as labours to returns are deemed to be included in the brickwork rates (SMM F10.C1F), and facing bricks have already been taken.
		Reveals	
2/	2.08	Closg. cavs., width: 50, slates 100 wide in c.m. (1:3), vert.	Closing the cavity is a linear item, stating the width of cavity and method of closing and plane (SMM F10.12.1.1.0). An alternative item would be 'Closing cavities, width: 50, single layer of hessian based bit. felt to BS 743 ref. A, lapd. 100 at jts. & b.&p. in c.m. (1:3), vertical.'
2/	2.08	Pla. to walls, width ≤ 300, in 2cts. a.b., bwk., trowld. fin. (reveals	Plasterwork ≤ 300 mm wide is so described and measured in metres (SMM M20.1.2.1.0).
	0.90	Ditto. conc., do. (soff.	Plasterwork to sides and soffits of openings is included with the abutting walls (SMM M20.D5), and work on different bases is kept separate (SMM M20.S5).
	5.06	Rdd. ∠ to pla., rad. 10 – 100.	Rounded angles to plaster are measured in metres as SMM M20.16, where in the radius range 10 – 100 mm. A fair joint of the plaster to the door frame is not measured, as it is deemed to be included in the plasterwork rates (SMM M20.C1a).
2/	0.08 2.08	Paintg. gen. surfs., prep. & 2ce emulsn. pt. pla. walls. (reveals	The painting to the reveals and soffit is included with the abutting walls as the finish is the same. It is not taken as a linear item as ≤ 300 mm girth as it forms part of the adjoining wall paintwork and does not therefore constitute an isolated surface.
	0.90 0.08	(soff.	

21.4

EXTERNAL DOOR (Contd.)

		Threshold len.	
		900	
		add ends 2/102 204	
		1.104	
1.10		Art. st. sill 300 x 150 x 1.104 lg. to BS 1217 grvd., fin. fair to 2nr. faces & 2 nr. ends & b.& p. in c.m. (1:3).	The threshold is a linear item classified as a sill with a dimensioned description as SMM F22.7.1.
		Below dpc	
0.90 0.15		Ddt. Bk. wall, thickness : 102.5, comms. in c.m. a.b. & Ddt. Ditto. facewk. o.s. in c.m. a.b. & Ddt. Form cav. to holl. wall, width : 50, a.b.	Deduction of brickwork for the threshold, with no allowance for the ends. The brickwork is built in cement mortar as it is below damp-proof course. Alternatively, the deduction of wall for the threshold could be taken with the door opening. Brickwork is only deducted for the full half-brick beds displaced (SMM F10. M3). It is assumed that adjustments of concrete bed, flooring and skirtings in the door opening have been taken elsewhere.

12 Measurement of Staircases and Fittings

TIMBER STAIRCASES

Most Part I examination measurement syllabuses include the measurement of relatively simple timber staircases. It is important to adopt a logical order of taking off in order to simplify the measuring process and reduce the possibility of omitting any items.

A good order to follow is

(1) Staircase complete as a composite item.
(2) Painting.
(3) Associated items such as sloping ceilings and cupboards under the stairs.

Timber staircases are enumerated as composite items with the type stated and supported by a dimensioned description or component drawing (SMM L30.1.1–2.0.0). Landings are included where they are part of the composite item.

It is sometimes necessary to calculate the amount of going or rise, by dividing the total length of a flight of stairs or the height between floors by the number of steps involved, where these are not shown on the relevant drawings. Part of a component detail of a tread and riser, illustrating the connecting joints is shown in Fig. XVIII.

Treads and risers are normally tongued and grooved and blocked together with their ends housed into strings. The bottom step may incorporate a semi-circular end. Wider staircases of 1.00 m or more will normally include rough timber carriages under the stairs with brackets giving support to the treads. Landings, strings and cappings, balusters or timber balustrades, together with attached handrails, where of the same materials as the staircase, newels or half-newels, and any spandril framing and panelling and soffit lining, where part of the component, are all included in a single enumerated composite item (SMM L30.C1–2). Newels are normally morticed and draw-bored to receive strings and morticed for handrails in accordance with BS 585. Isolated handrails and balustrades which do not form an integral part of a staircase unit are measured separately in metres with a dimensioned description.

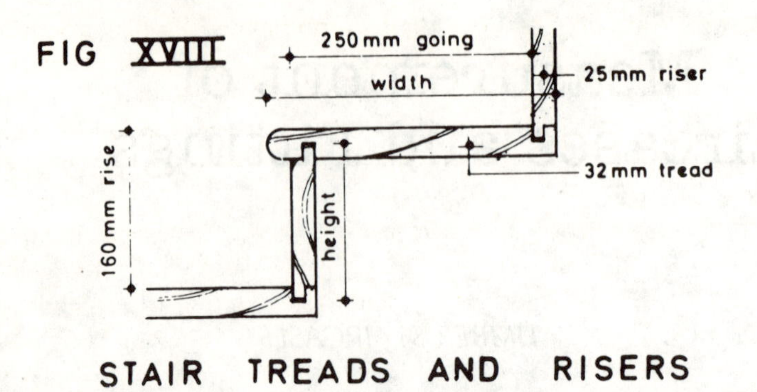

STAIR TREADS AND RISERS

It is essential to make reference in the staircase description to the appropriate component drawing which should accompany the bill for tendering purposes, and the drawing must contain all the information necessary for the manufacture and assembly of the staircase (SMM General Rules 5.2). Detailed descriptive information is not required in the bill where a component drawing is supplied, since the purpose of the billed item will then be to identify the staircase with the relevant component drawing.

Where a component drawing is not provided, the work will be covered by an extensive dimensioned description as SMM L30.1.1.0.0.

Painting of staircase components will follow the enumerated composite item. Painting of skirtings, strings, cappings, handrails, newels, balusters, and margins of treads and risers will all be classified as general surfaces and will be measured in metres where $\leq$ 300 mm girth and in m^2 when $>$ 300 mm girth (SMM M60.1.0.1–2.0).

Plasterwork and painting to walls and ceilings in staircase areas are kept separate because of the restricted working space (see SMM M20.M3 and M60.M1).

FITTINGS

Measurement of joinery fittings, such as cupboards, work-benches, counters, dressers, bookcases, and the like, are also taken as enumerated composite joinery items. For example a kitchen fitment item might be described as follows:

> Supply kitchen fitment in wrought softwood, 800 mm wide × 1800 mm high ×400 mm deep, overall, consisting of 3 nr cupboards, 4 nr drawers and a pull-down table top, all in accordance with component detail drawing 'X' (fixing measured separately).

Alternatively, the supply of the fitting by a nominated supplier could be the subject of a prime cost sum as SMM A52.1.1–2.1.0, including a description similar to that already given and with an accompanying item for the inclusion of the main contractor's profit. Fixing is measured with the appropriate work section, where prime cost sums are grouped together in the Bill.

Where the exposed woodwork is to be stained or polished, the description of the timber shall state that it is to be selected and protected for subsequent treatment (SMM L30.S4), and this can often, with advantage, be included in the heading.

Stock-pattern cupboards and other fitments are enumerated giving the appropriate supplier's catalogue reference as SMM General Rules 6.1, together with the separate enumerated fixing item as described previously.

Isolated shelves, worktops and similar features are measured in metres, giving a dimensioned overall cross-section description as SMM P20.3.1.0.0. All labours will be included in the descriptions of these items, including any enumerated stopped labours. The work is deemed to include ends, angles, mitres, intersections and the like, except on hardwood items $> 0.003\ m^2$ in sectional area (SMM P20.C1).

WORKED EXAMPLE

A worked example follows covering the measurement of a simple timber staircase and shelving to a larder.

TIMBER STAIRCASE AND SHELVING TO LARDER   Drawing No. 17

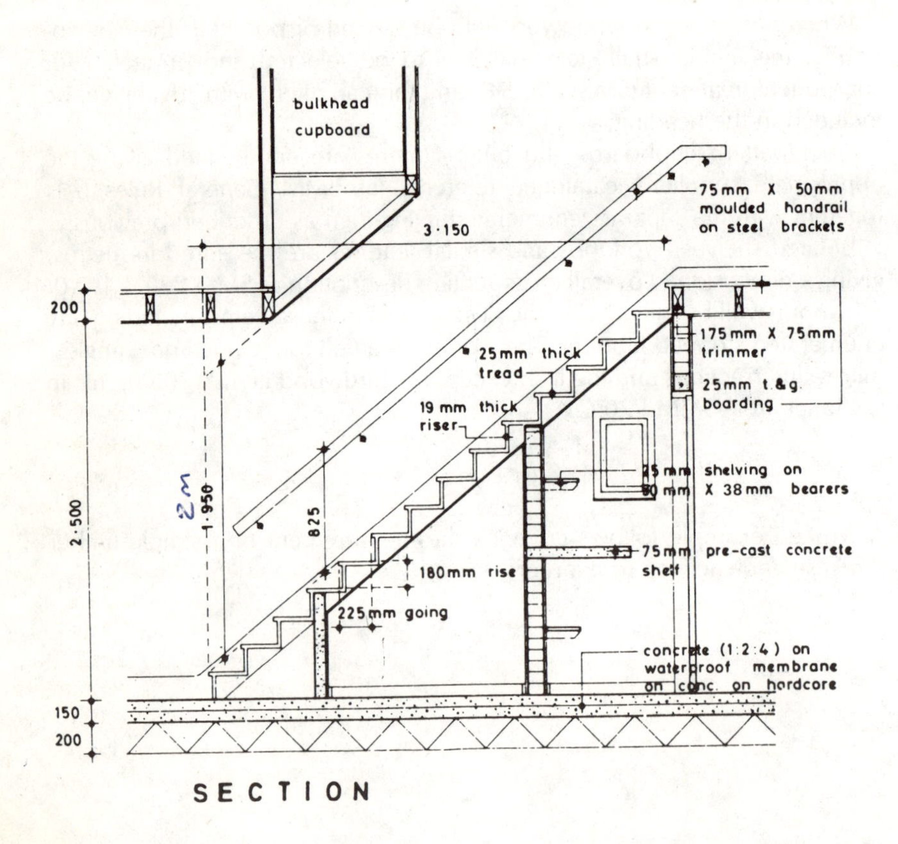

SECTION

PLAN

scale 1:50

TIMBER STAIRCASE

EXAMPLE XXII

	1		Wrot swd. single flight tbr. staircase as component drwg. X.
			Wall strgs.
2/	4.50		Paintg. gen. isoltd. surfs., wd., gth. ≤ 300, k.p.s. & (3).
			Hdrl. 4.000 / 300 / 4.300
	4.30		Isoltd. hdrl. 50x75 wrot Honduras Mahogany mo., jtd. w. hdrl. scrs., sel. & protectd. for subseqt. treatmnt., fxd. to stl. bkts. (m/s).
			&
			Paintg. gen. isoltd. surfs., hwd., gth. ≤ 300, prep. 2ce oil & w.p., in staircase area.

SMM L30.1.1-2 states that staircases shall be enumerated and supported by a component drawing or dimensioned description. SMM General Rules 5.2 makes it clear that the drawings must contain all the information necessary for the manufacture and assembly of the staircase. These details will include all the joints, labours and fixings, as illustrated in the staircase details shown in 'Building Technology' by the same author. If a component drawing is not available, then the work is covered by a dimensioned description, which is likely to be extremely lengthy. When a component drawing is used no descriptive information is necessary, only a reference to the drawing(s).

The length of the strings has been scaled from the drawing. Painting is taken as to isolated general surfaces in metres, where the total girth is ≤ 300 mm (SMM M60.1.0.2.0).
The handrail has been measured as a separate linear item as SMM P20.7.1.0.0 with a dimensioned overall cross-section description, as it does not form an integral part of the staircase. If however it was supported by balusters or a balustrade, then it would form part of the staircase, although if of a different material to the balustrade, it would be measured separately as a linear item classified as associated handrails (SMM L30.3.1.0.0).
The description of the handrail is to include the method of jointing and also the method of fixing, where not at the discretion of the contractor (SMM P20.S7-8).
Selection and protection of the timber for subsequent treatment is included in the description (SMM P20.S4).
Painting to handrails is classified as to general isolated surfaces (SMM M60.1.0.2.0), including reference to staircase area as SMM M60.M1.

TIMBER STAIRCASE (Contd.)

Timesing	Dimension	Description	Notes
2/	1	E.O. hwd. hdrl. for end.	Ends and mitres are enumerated as extra over the handrail, as the sectional area > 0.003 m^3 (SMM P20.C1).
	1	E.O. hwd. hdrl. for mitre	
6/	1	Fix only stainless stl. hdrl. bkt., scrd. to hwd. & plugd. & scrd. to bwk.	Handrail brackets are taken as enumerated fixing items. The supply of the brackets would probably be included in a prime cost sum covering all ironmongery.
		Pla. soff. to strs.	
	1.75 0.90 1.15 0.90	Pla. clg., pla. bd. 10th. nailed to swd. w. skim ct. 3th. a.b. in staircase area. & Paintg. gen. surfs., prep. & 2ce emulsn. pt. on pla. clg. in staircase area.	Plasterboard is measured in m^2 giving the particulars listed in SMM M20.2.1.2.0 (See Example XVI). Work in staircase areas is to be given separately and so described (SMM M20.M3). The purpose of keeping separate work in staircase areas, including surrounding walls or ceilings, is to allow the estimator to price working off staircase flights and/or in restricted spaces. Decorations to staircase areas are classified separately (SMM M60.M1). Note: The brick balustrade wall will probably be measured in brick/block walling. Whilst as to finishes, plaster to vertical surfaces is measured in m^2, with plaster to the top of the balustrade wall taken as a separate linear item and with linear items for the external angles if rounded with a radius in the 10 – 100 mm range (SMM M20.16). Painting to newels, balusters and timber balustrades would all be classified as to general surfaces in the same way as the margins to treads and risers (SMM M60.1.0.2.0) An alternative approach would be to classify all these items as to 'staircase areas' as SMM M60.M1.
		Paintg. to margins o/a. gth. of t.&r. going 225 nosg. 20 riser 180 425	
14/2/ 2/	0.43 0.18	Paintg. gen. isoltd. surfs., wd., gth. ≤ 300, k.p.s. & ③., in staircase area. (margins to t.&r.) (top riser to 1st.flr.)	

TIMBER STAIRCASE (Contd.)

		Description	Side notes
		Shelvg. to larder	
2/	0·90	Isoltd. shelves 25 x 225 wrot swd. on 38 x 50 sn. swd. brrs. plugd. to bwk.	Isolated shelves are measured in metres giving a dimensioned overall cross-section description as SMM P20 3.1.0.0, and including the method of fixing as SMM P20.S8.
		&	
		Paintg. gen. isoltd. surfs., wd., gth. ≤ 300, k.p.s. & ③.	Painting of shelf edges is taken as a linear item as SMM M60.1.0.2.0.
		Conc. shelvg.	
		len. 900, width 700; add beargs. 2/100 200; 1·100	Note: deduction of brickwork would have to be made for the volume occupied by the shelf in accordance with SMM F10. M3.
	1	Precast conc. shelf (1:2:4/20 agg.) 1100 x 700 x 75th. w. smth. top & 1nr. fair edge, bedded in c.m. (1:3).	The shelf is enumerated with a dimensioned description as SMM F31.1.1.0.0, and giving the particulars listed in SMM F31.S1-5. Moulds are deemed to be included with precast items (SMM F31.C1).

22.3

13 Measurement of Plumbing Installations

ORDER OF MEASUREMENT

As with the measurement of most other types of building work, it is important to adopt a logical order of taking off. The following order is frequently followed in practice.

(1) Connection to water authority's main and all work up to boundary of site, including reinstatement of public highway and provision of stop-valve near the site boundary.
(2) Underground service and rising main from site boundary up to cold water storage tank, including any stopvalves, holes through walls, ceilings and floors, lagging of pipes, and the like.
(3) Branches to rising main, such as supply to sink, including any associated work.
(4) Cold water storage tank or cistern and associated work, such as bearers, overflow, cover and insulating lining.
(5) Down services with branches, including any stopvalves, holes through walls, ceilings and floors and lagging of pipes.
(6) Sanitary appliances such as sinks, wash basins, baths and water closets – supply and fixing, including supporting brackets, taps and the like.
(7) Discharge pipes (waste, soil and vent pipes) and associated work.
(8) Any other work connected with the plumbing installation, such as painting pipes and testing the installation.

DRAWINGS OF PLUMBING INSTALLATIONS

The architect's drawings often give little information on the layout of the water service pipes in a building. The information supplied is frequently limited to the point of entry of the rising main into the building, the position of the cold water storage tank or cistern and the various sanitary appliances. In these circumstances the quantity surveyor must decide on a suitable layout of pipes and plot them on the various drawings before he can start the taking off work.

It is good policy to draw the various pipes in different colours, such as the rising main in red and down services in blue. The stopvalves required should also be indicated on the drawings, and it is good practice to provide one on each branch.

CONNECTION TO WATER MAIN

The tapping and insertion of a ferrule to the water authority's main and the provision of the communication pipe from the main to the boundary of the site with a stopvalve provided at this point, including opening up and reinstatement of the highway and all watching and lighting, is usually covered by a provisional sum based on the cost of previous projects or quotations from the water authority and local authority who will normally carry out the work (SMM A53.1.1–2.0.0).

PIPEWORK GENERALLY

Pipes are classified under appropriate headings, such as cold water system (S10 of *Common Arrangement*), and measured over all fittings and branches in metres, stating the type, nominal size, method of jointing and type, spacing and method of fixing supports, and distinguishing between straight and curved pipes (SMM Y10.1.1.1.0 and Y10.M3). Pipes are deemed to include joints in their running length (SMM Y10.C3). Furthermore, the provision of everything necessary for jointing is deemed to be included without the need for specific mention (SMM Y10.C1). The type of background to which the pipe supports are fixed will be classified in the categories listed in SMM General Rules 8.3.

Details of the kind and quality of materials used in the pipes, gauge and other relevant particulars listed in SMM Y10.S1–6, are likely to be included in preamble clauses or a project specification.

Made bends, special joints and connections, and fittings such as Y-junctions, reducers, elbows, tees and crosses, are all enumerated as items extra over the pipes in which they occur (SMM Y10.2.1–4). In the case of special joints, the type and method of jointing is to be stated and they comprise joints which differ from those generally occurring in the running length or are connections to pipes of a different profile or material, connections to existing pipes or to equipment, appliances or ends of flue pipes (SMM Y10.D2).

Pipe fittings $\leq$ 65 mm diameter are classified according to the number of ends, while those of larger diameter are described. The method of jointing is stated where different from the pipe in which the fitting occurs.

Valves and cocks are classified as pipework ancillaries and are enumerated, stating the type, nominal size, method of jointing, type, number and method of fixing supports and type of pipe to which connected (SMM

Y11.8.1.1.0). Those located in ducts or trenches are each kept separate and so described.

WATER STORAGE TANKS OR CISTERNS

Water storage tanks or cisterns are enumerated giving appropriate particulars as SMM Y21.1.1.0.0, or alternatively a cross-reference may be made to the project specification, which will substantially reduce the length of the item description. A typical example of a storage cistern description is given in example XXIII. Timber supports to tanks are measured in metres with a dimensioned overall cross-section description as SMM G20.13.0.1.0. Insulation to tanks or cisterns is enumerated, giving the overall size of the tank, or insulation contained in casings can be measured in m^2, in accordance with SMM Y50.1.4.1–2.0, and giving in both cases the particulars listed in SMM Y50.S1–5. Any pipework located in the roof space should be insulated and this insulation is measured in metres giving the type of insulation and nominal size of the pipe (SMM Y50.1.1.1.0). The insulation item is deemed to include smoothing the materials, working around supports, pipe flanges and fittings, excluding metal clad facing insulants (SMM Y50.C1).

HOLES FOR PIPES

Cutting holes through the structure for pipes and making good surfaces are enumerated, stating the nature and thickness of the structure and the shape of the hole, and classifying the pipes as to size in accordance with SMM P31.20.2.1–3.2 & 4; for example, pipes $\leq$ 55 mm nominal size, 55–110 mm and $>$ 110 mm. The cutting of holes for pipes is best picked up when the various lengths of pipework are being taken off, rather than leaving all the holes to be taken off after the pipework has been measured complete. By contrast, painting of pipes may often, with advantage, be left to the end of the taking off.

SANITARY APPLIANCES

Sanitary appliances include low level WC suites, WC pans and cisterns, urinals and cisterns, sinks which are not supplied as part of a kitchen fitting installation, wash basins, bidets, baths with bath panels and trim, showers and vanity units (*Common Arrangement N13*). These appliances are enumerated giving details of the type, size, pattern and capacity, as appropriate, and method of fixing, and may include a cross-reference to the project specification (SMM N13.4.1.1.6). The supply of appliances is frequently covered by a prime cost sum, including relevant particulars and provision for main contractor's profit as SMM A52.1.1.1.0 with separate enumerated fixing items. Another alternative is to give supply and fix items with full details including the supplier's catalogue reference numbers (SMM General Rules 6.1).

BUILDER'S WORK IN CONNECTION WITH PLUMBING INSTALLATION

Builder's work in connection with a plumbing installation is identified under an appropriate heading (SMM P31.M2). Unless identified in SMM work sections P30 and P31, all other items of builder's work associated with the plumbing installation are given in accordance with the appropriate work sections (SMM P31.M1). Where the hot water and heating installation is to be carried out by a nominated sub-contractor, items will be provided to cover any specific items of special attendance required in accordance with SMM A51.1.3.1–8.1–2, classified as either fixed or time-related charges. General attendance on nominated sub-contractors is measured in accordance with SMM A42.1.16.

WORKED EXAMPLE

A worked example follows covering the measurement of a cold water supply system, sanitary appliances and discharge pipes.

PLUMBING INSTALLATION

Drawing No. 18

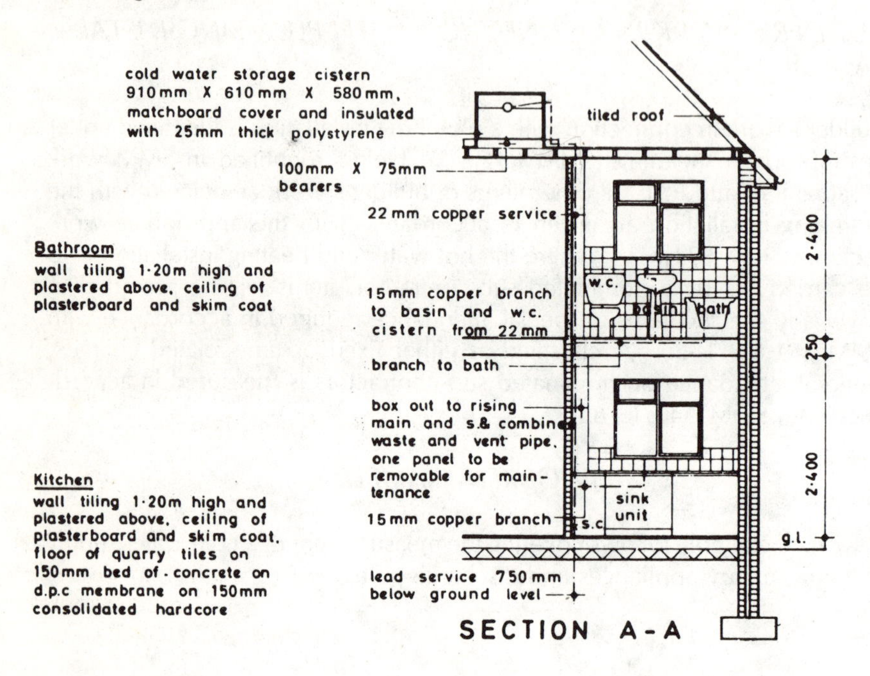

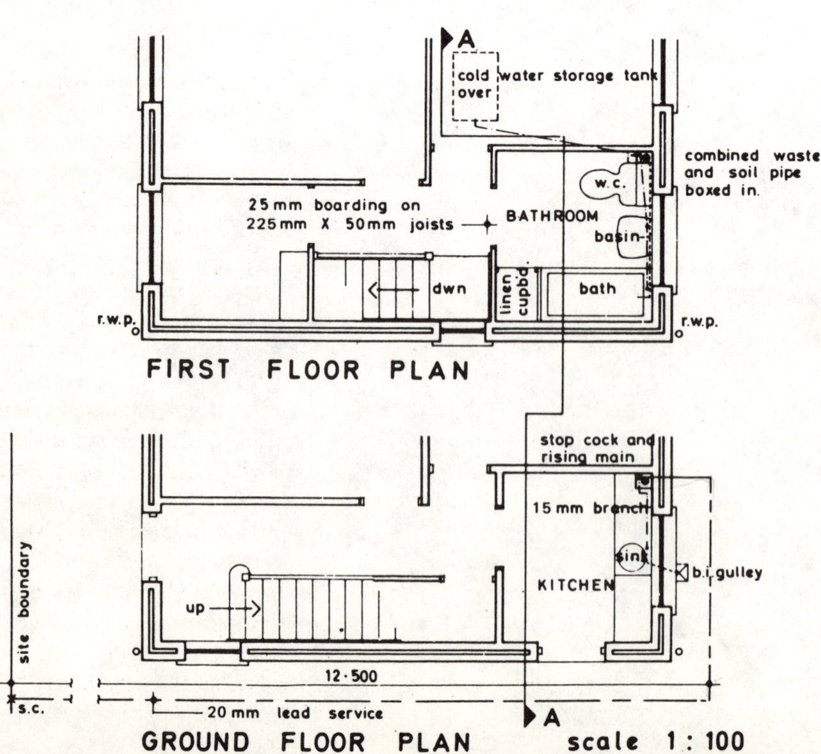

PLUMBING	INSTALLATION		EXAMPLE XXIII
		COLD WATER SUPPLY SYSTEM Conn. to Main.	Heading as Common Arrangement S10; this avoids the need to describe the individual items that follow as cold water services.
		Allow the Provsnl. Sum of £150 for conn. to main in rd. & bringg. 22 ld. serv. up to bdg. of site, inc. provsn. of stop valve & box by Water Authority.	Work to be carried out by the water authority and highway authority are covered by provisional sums as SMM A53.1.1-2.0.0.
		Allow the Provsnl. Sum of £60 for perm. reinstatement of the highway by the Highway Authority.	
		Cold Water Service	Lead service pipes are no longer permitted under the Water Supply Byelaws 1986, but some are incorporated to show the method of measuring and describing them.
	1	E.O. ld. pipe, nom. size 22, for spec. conn. s.j. to xtg. fittg. (water authority S.V.	This joint is classified as a special connection as SMM Y10.2.2.1.0, and requires a separate enumerated item as the stop valve is not part of the measured work and is regarded as existing (SMM Y10.D2).
		12·500 3·000 550 16·050	
	16·05	Pipes, st., nom. size: 22, ld. to BS 602, Table 1, code 11, w. sold. jts. in trs. & Exc. tr. for serv., nom. size ≤ 200, av. depth: 500-750.	The service pipe is measured in metres stating the type, nominal size and method of jointing and including laying the pipe in trenches as SMM Y10.1.1.1.3. Any joints in the running length are deemed to be included (SMM Y10.C3). Excavation of trenches for service pipes is measured in metres, grouping all services together which are ≤ 200 mm nominal size and giving the average depth in 250 mm stages (SMM P30.1.1.1-2). Excavating trenches is deemed to include earthwork support, consolidation of trench bottoms, trimming excavations, special protection of services, backfilling and compaction of excavated materials and disposal of surplus excavated materials (SMM P30.C1). Hence these items do not require inclusion in the trench excavation description.
	2	E.O. ld. pipe nom. size: 22 for made bend.	Made bends are enumerated as extra over the pipe in which they occur (SMM Y10.2.1).

23.1

PLUMBING INSTALLATION (Contd.)

Timesing	Dimension	Squaring	Description	Waste	Side notes
			thro' wall 300 up to g.l.s.v. 1.050 1.350		Build up of length of service pipe from outside the building up to the first internal stop valve.
	1.35		Pipes, st., nom. size: 22, ld., a.b.d., w. sold. jts. & fxg. w. tinned stl. saddle clips @ 1.2 m ccs. to masonry.		Pipes are all measured in metres stating the type, nominal size, method of jointing and supports (SMM Y10.1.1.1.1). The use of letters a.b.d. (as before described) avoids the need to repeat the class and code of pipe.
	1		E.o. ld. pipe nom. size: 22 for made bend.		Made bends are described and enumerated as extra over the pipes in which they occur (SMM Y10.2.1).
			Rising Main		
	1		Pipewk. ancillary, nom. size: 22, brass h.p. comb. scrdn. stop valve & draw off tap to BS 1010 w. 1 nr. s.j. to ld. & 1 nr. jt. to cop. pipe.		Valves are enumerated and described as ancillaries in accordance with SMM Y11.8.1.1.0.
			len. g. flr. sty. ht. 2.250 f. flr. jsts. & bdg. 250 f. flr. sty. ht. 2.400 clg. joists 150 5.050		Build up of length of rising main in waste. The majority of pipe lengths in this example are obtained by scaling, although preference should always be given to figured dimensions where available.
	5.05		Pipes, st., nom. size: 22 cop. to BS 2871, Pt. 1, Table X, w. cap. jts. to BS 864, Pt. 2, fxd. w. cop. clips plugd. & scrd. to masonry @ 1.50 m ccs.		Description of pipe as SMM Y10.1.1.1.1, including jointing and fixing. The background to which it is fixed should be classified in accordance with SMM General Rules 8.3.
			across rf. space 2.300 up to inlet of cistn. 600 2.900		A distinction is made between pipes which are fixed to timber, masonry (concrete, brick, block and stone) and vulnerable materials (glass, tiled finishes and the like).
	2.90		Pipes, st., nom. size: 22, cop. a.b.d. but fxd. w. cop. clips scrd. to tbr. @ 1.50 m ccs. & Insulatn., 'Armaflex' foamed plastic, 10th. & bindg. w. cop. wire to pipes, nom. size: 22.		Pipe insulation is measured in metres, stating the type of insulation and nominal size of pipe (SMM Y50.1.1.1.0), and the kind and quality, thickness, finish and method of fixing of insulation (SMM Y50.S1-5).

23.2

PLUMBING INSTALLATION (Contd.)

Timesing	Dimension	Squaring	Description	Notes
	1		Cut circ. hole thro. 225 bk. holl. wall for pipe ≤ 55ϕ & m/gd. & Cut circ. hole in 150 conc. flr. & polyth. w.p. memb. for pipe ≤ 55ϕ & m/gd. & Cut circ. hole thro.' q. tile pavg. for pipe ≤ 55ϕ & m/gd. & Cut circ. hole thro. 25 swd. for pipe ≤ 55ϕ.	Enumerated item for hole giving the nature of the structure, pipe size classification and other relevant particulars as SMM P31. 20.2.1–3.4. If pipe sleeves are specified, these will be enumerated, and the description will include fixing, type, size of pipe and nature of structure (SMM P31.23.2.1.6).
2/	1		Cut circ. hole thro. plabd. 10th. & pla. skim ct. 3th. for pipe ≤ 55ϕ & m/gd.	
2/	1		E.o. cop. pipe, nom. size: 22, for made bend.	Enumerated item as SMM Y10.2.1.
	1		Pipewk. ancillary: ball valve, nom. size: 22, h.p. diaphragm type (brass body) to BS 1212, Pt. 2, w. brass lever & PVC float to BS 1968, inc. strt. tk. conn. & strt. cplg. & comp. jt. to cop. pipe.	The ball valve is enumerated as an ancillary item in accordance with SMM Y11.8.1.1.0, giving all necessary particulars in the description, including jointing, fixing and type of pipe.
			<u>Branches</u> <u>Sink</u>	Pipe fittings are enumerated as extra over the pipes in which they occur. Cutting and Jointing pipes is deemed to be included (SMM Y10.C5).
	1		E.o. cop. pipe, nom. size: 22, for fittg., 3 nr. ends. (reducg. tee	In the case of reducing fittings, the largest diameter only is stated (SMM Y10. M5). Fittings to pipes ≤ 65 mm diameter are described as fittings, stating the number of ends (SMM Y10.2.3.2–5). An abbreviated description can be used as similar items have been taken previously.
	1.35		Pipes, st., nom. size: 15, cop. & fxg. to masonry a.b.d. 23.3	

PLUMBING INSTALLATION (Contd.)

Timesing	Dimension	Squaring	Description	Side notes
	2		E.O. cop. pipe, nom. size: 22, for made bend.	These bends are needed to secure alignment with the tap connectors.
	1		Pipewk. ancillary: pillar tap, nom. size: 15, chrom. plated easy cln. patt. high necked to BS 1010 indexed COLD inc. strt. tap. conn. & comp. jt. to cop. pipe.	The tap will form a separate enumerated item as it is not usually provided with the sink. In practice a pair of taps would normally be required (one for cold water and the other for hot).
	1		Pipewk. ancillary: stop valve, nom. size: 15 brass h.p. scrdn. type to BS 1010 w. 2nr. comp. jts. to cop. pipe.	It is good practice to provide a stop-valve on each branch even if not shown on the drawing.
	1		Cut circ. hole thro. 19 swd. for pipe ≤ 55 ϕ. (side of sink unit	Enumerated cutting hole item as SMM P31.20.2.1.2.
			Bath	
	1		E.O. cop. pipe, nom. size: 22, for fittg., 3nr. ends. (tee	Enumerated fitting item as SMM Y10.2.3.4.0.
	1.75		Pipes, st., nom. size: 22, cop. & fxg. to masonry a.b.d. (hor. serv.	
	0.70		(vert. serv. to bath	
2/	1		E.O. cop. pipe, nom. size: 22, for made bend.	Note: all fittings are supplied from the rising main in this example, to keep it relatively simple, although in many instances a separate low pressure down service will be provided.
	1		E.O. cop. pipe, nom. size: 22, for spec. conn. to pillar taps (m/s) inc. bent connector & comp. jt.	Branch services to fittings consist of a length of branch pipe, connection to main pipe and a connection to the fitting.
	1		Pipewk. ancillary: stopvalve, nom. size: 22, brass hp scrdn. to BS 1010 w. 2nr. comp. jts. to cop. pipe.	In addition a stop valve may be inserted on the branch and bends are needed at any change of direction. An enumerated special connection item in accordance with SMM Y10.2.2.1.0 has been measured in connection with the pillar tap.

23.4

PLUMBING INSTALLATION (Contd.)

Timesing	Dimension	Squaring	Description	Notes
			Wash basin	Enumerated extra over fitting items are deemed to include cutting and jointing pipes to fittings (SMM Y10. C5). If it was required to paint the pipe, the painting item would follow the pipe.
	1		E.o. cop. pipe, nom. size: 22, for fittg., 3 nr. ends. (reducg. tee	
	1·15		Pipes, st., nom. size: 15, cop. & fxg. to masonry a.b.d.	
	1		E.o. cop. pipe, nom. size: 15 for made bend.	Made bends are enumerated as extra over the pipe (SMM Y10. 2. 1).
	1		E.O. cop. pipe, nom. size: 15 for spec. conn. to pillar tap (m/s) inc. bent connector & comp. jt.	Enumerated special connection item as SMM Y10. 2. 2. 1. 0. Nominal size of connection stated if different from pipe in which joint or connection occurs.
			WC	
	0·90		Pipes, st., nom. size: 15, cop. & fxg. to masonry a.b.d.	It is necessary to adopt a logical sequence of taking off to avoid the omission of any items. In this example each main pipe run or branch has been taken separately, starting with the first fitting and following through with the pipe, fittings, labours and finally any holes in the structure of the building.
	1		E.O. cop. pipe, nom. size: 15 for spec. conn. to cistn., inc. bent connector & comp. jt.	Some surveyors might prefer to adopt the slightly different sequence of pipes, fittings, ancillaries and builder's work in connection with the particular service pipe or branch. Builder's work can either be taken with the plumbing or dealt with separately.
			Storage cistn.	
	1		Cistn. galvd. m.s. & cover to BS 417, grade B, ref. SC70, of 227 litres cap. & o/a size 910 x 610 x 580, w. perfs. for 2 nr. pipes & apply intly. 2 cts. of bit. paint.	The cistern is an enumerated item, giving the type, size and capacity as SMM Y21. 1. 1. 0. 0. Note the use of British Standards in descriptions and the need to give grades and other classifications where appropriate.
2/	1.00		Indvidl. suppts. 75 x 100 sn. swd.	The cistern bearers have been measured in metres as individual supports, giving a dimensioned overall cross-section description as SMM G20. 13. 0. 1. 0.

23.5

PLUMBING INSTALLATION (Contd.)

Dimension	Description	Notes
1	Insulatn. polyst. 25th. as casg. to sides & top of cistn, 910 x 610 x 580 o/a size, & tyg. w. cop. wire.	Insulation to equipment can be enumerated, stating the overall size of the equipment or taken in m^2 (SMM Y50.1.4.1-2).
	Overflow	
2·20	Pipes, st., nom. size: 28, cop & fxg. to tbr. a.b.d.	Measured in metres as SMM Y.10. 1.1.1.1.
1	E.O. cop. pipe, nom. size: 28, for spec. conn. to cistern inc. flgd. tk. conn. & strt. conn. w. comp. jt.	Connection of overflow pipe to cistern is an enumerated item in accordance with SMM Y10.2.2.1.0, but with no requirement to give the nominal size of the connection unless it differs from the pipe.
1	Cut circ. hole thro. rf. tilg. for pipe ≤ 55 ϕ. & F.O. lead slate, 300 x 300, w. collar ard. 28 ϕ pipe.	Holes for pipes in roof coverings are enumerated. The fixing of soakers and saddles (provided by other trades) are enumerated separately, with a dimensioned description (SMM H60. 10.6.1.1). A similar approach has been adopted for the lead slate.
1	Lead slate Nr. 5 300 x 300 in size w. collar & soldg. to 28 ϕ cop. pipe (handed to others for fixg.).	Metal slates (for fixing by slater or tiler) are enumerated, with a dimensioned description (SMM H70. 26.1.0.1).
	SANITARY APPLIANCES AND ASSOCIATED FOUL DRAINAGE WORK ABOVE GROUND	Sanitary appliances are covered in SMM Section N13 and foul drainage above ground in Section R.11.
	San. appliances	
	Provide the P.C. Sum of £400 for san. appliances, comprisg. comb. stainless stl. drainer & sk., bath, ped. wash basin & low lev. wc suite, delvd. to site. & Add for profit.	Supply only appliances is covered by a prime cost sum with provision for the addition of profit as SMM A52.1.1.1.0. The prime cost item is to include a description of the components covered by the stated sum.

23.6

PLUMBING INSTALLATION (Contd.)

Assemble & fix the following sanitary appliances (jts. to pipes m/s).

Timesing	Dimension	Squaring	Description	Side notes
	1		Combined drainer & bowl, stainless stl. drainer 1067 x 467 & rect. bowl 457 x 343 x 178 dp., w. waste fittg. chain, stay & plug & fxg. to sink unit (m/s).	The stainless steel sink top is fixed to the top of the sink unit. The sink unit forms a separate measured item.
	1		Bath, porcelain enam. cast iron 1700 lg. to BS 1189 w. 2nr. chr. pltd. pillar taps, waste fittg., chain, stay & plug.	Bath panel and bearers are measured separately. The bath trap will be taken with the wastes.
	1		Wash basin ped. in glzd. f/clay 559 x 406, w. 2nr. chr. pltd. pillar taps & waste fitting, chain, stay & plug & fxg. on m.s. angle bkts. p & s. to masonry & scrg. ped. to swd.	An adequate description of each appliance is needed to enable the contractor to satisfactorily compute a price for assembling and fixing.
	1		WC suite, low lev. w. vit. china ped, dble plastic ring seat, 9 litre glzd. flushg. cistn on pr. of galvd. m.s. concld. bkts., enam. flush pipe, inc. scrg. ped. to swd., jtg. outlet to ld. discharge pipe (m/s), p.&s. cistn. bkts. to masonry & jtg. fl. pipe to cistn. & arm of ped.	The WC pan, flushing cistern and flush pipe are combined in a single enumerated item.
			(End of assembly & fix)	It is advisable to indicate the end of the fixing items to avoid any confusion.
			Bath	
	1		Vitrolite panel to bath side, 1700 x 500 fxd. w. chr. pltd. rd. hdd. scrs. to swd. suppts. (m/s).	A panel is required to one side only of the bath. The bath panel bearers are measured in metres as framed supports giving a dimensioned overall cross-section description and spacing of the members as SMM G20.12.2.1.0. Method of fixing of members is only described where not at the discretion of the contractor (SMM G20. S2).
2/	1.70		Framed suppts., 40 x 50 sn. swd., spaced 500 apart.	
4/	0.50			

23.7

PLUMBING INSTALLATION (Contd.)

Quantity	Description
	WC overflow
0.60	Pipes, st., nom. size: 22, cop. & fxg. to tbr. a.b.d.
1	E.O. cop. pipe, nom, size: 22 for spec. conn. to flushg. cistn. inc. bent connector & comp. jt.
1	E.O. cop. pipe, nom. size: 22 for fittg., hinged cop flap & fr. to splay cut end w. sold. jt. & Cut circ. hole thro. 255 bk. holl. wall for pipe ≤ 55 ϕ & m/gd. & Cut circ. hole thro. wall tilg. for pipe ≤ 55 ϕ & m/gd.
	Waste Pipes
1	Pipewk, ancillary: trap, nom. size: 32 drawn cop. 'S' type w. 38 seal to BS 1184, w. scrd. jt. to waste o/let on appliance & comp. jt. to cop. pipe. (wash basin
1	Pipewk, ancillary: trap, nom. size: 38 drawn cop. 'S' type w. 38 seal to BS 1184, w. scrd. jt. to waste o/let on appliance & comp. jt. to cop. pipe. (sink
1	Pipewk. ancillary: trap, nom. size: 38 brass w. 88½° outlet shallow seal & cleaning plug ea. side & w. scrd. jt. to waste o/let on appliance & comp. jt. to cop. pipe. (bath

23.8

Framed supports are where the members are jointed together other than butt jointed (SMM G20.D8).
Note use of various sub-headings to act as signposts thoughout the dimensions.
Alternatively all overflow pipes could be grouped together in the taking off.
An enumerated item to cover the connection of the pipe to the cistern, which includes a bent connector to obtain the desired direction (SMM Y10.2.2.1.0).
This could be classed as a pipe fitting under SMM Y10.2.3.5. The flaps now seem rarely to be used in practice.
Two enumerated items to cover the overflow pipe passing through the hollow wall and internal wall tiling in accordance with SMM P31.20.2.1.2 and 4. If a pipe sleeve was required to accommodate the pipe where it passes through the wall, this would be enumerated, stating the type, size of pipe and nature of structure as SMM P31.23.1.1.5, such as 'galvanised mild steel pipe sleeve for 22 pipe, bedded and pointed in prepared hole in 255 hollow brick wall.'
Traps could be classified as ancillaries and enumerated in accordance with SMM Y11.8.1.1.0.

The description is to include the type, nominal size, method of jointing and supports, and type of pipe (SMM Y11.8.1.1.0).

The bath trap is in brass with a shallow seal because of the restricted space available.

PLUMBING INSTALLATION (Contd.)

	2.00 1.80		Pipes, st., nom. size: 38, cop. w. cap. jts. & fxg. to masonry a.b.d. (bath (sink	Pipes are measured in accordance with SMM Y10.1.1.1.1, including any joints in running lengths (SMM Y10.C3).
	0.80		Pipes, st., nom. size: 32, ditto. (wash basin	Waste from wash basin is connected into the bath waste.
2/	2		E.O. cop. pipe, nom. size: 38, for made bend. (sink & bath	Made bends are enumerated as extra over pipes (SMM Y10.2.1) - taking 2 bends to each length of waste pipe.
	1		E.O. cop. pipe, nom. size: 38, for fittg., 3nr. ends. (reducg. tee (wash basin & Cut circ. hole thro. 255 bk. holl. wall for pipe ≤ 55ϕ & m/gd. masonry. (sink & E.O. cop. pipe, nom. size: 38, for spec. conn. to clay back inlet gully, w. c.m. (1:2) jt.	The reducing tee is an enumerated fitting (extra over pipe) item as SMM Y10.2.3.3.0, stating the largest pipe size as SMM Y10.M5. Enumerated item as SMM P31.20.2.1.2&4. Connection to gully below grating level through a back inlet. Section R11 of SMM covers drainage above ground.
		FOUL	DRAINAGE ABOVE GROUND Discharge Pipe len. above rf. 1.000 gl. to rf. 5.200 6.200	The term discharge pipe has been used in BS 5572 (Code of Practice for Sanitary Pipework) in substitution for soil and waste pipe, and this practice is followed in the Building Regulations 1985.
	6.20		Pipes, st., nom. size: 102, s&s c.i. to BS416, type A, w. caulked ld. jts. & fxg. w. galvd. stl. hinged holderbats into masonry @ 1 m ccs.	The length is taken from 1.00 m above roof covering (assuming a hipped end) down to the upper surface of the concrete slab to the ground floor where it will be jointed to a clay pipe. Note the use of British Standards in the descriptions of pipes. The description of the pipe is to include the type, spacing and method of fixing supports.
22/7/	1.00 0.10		Paintg. servs., met., ext., prep. prime & ③.	Painting to the pipe above roof level is measured in m² as it > 300 mm girth (SMM M60.9.0.1.0). Holderbats are deemed to be included (SMM M60.C8). External work is so described (SMM M60.D1).

23.9

PLUMBING INSTALLATIONS (Contd.)

Timesing	Dimension	Description	Side notes
	1	E.O. c.i. pipe, nom. size: 102, for spec. conn. to clay pipe in c.m. (1:2). (jt. between bott. of disch. pipe & drain) &	The connection is an enumerated item as SMM Y10.2.2.1.0.
		Pipewk. ancillary: gtg. to outlet of 102 c.i. pipe, galvd. wire balloon to BS 416.	Balloon gratings are enumerated stating the type, nominal size and type of pipe (SMM R11.6.4.1.0).
	1	Cut circ. hole thro. rf. tilg. for pipe 55 – 110 ϕ. &	Hole enumerated as SMM P31.20.2.2.2. and classified in the 55-110 mm pipe diameter category.
		F.O. lead slate 450 x 450 w. collar ard. 102 ϕ pipe. &	Separate fixing item as SMM H60.10.6.1.1, including a dimensioned description.
		Lead slate Nr. 5, 450 x 450 in size w. collar turned into soc. of 102 c.i. pipe (handed to others for fixg.).	Enumerated item for supplying lead slate with a dimensioned description (SMM H70.26.1.0.1).
2/	1	Cut circ. hole thro. pla. bd. 10th. & pla. skim ct. 3th. for pipe 55 – 110 ϕ & m/gd.	As SMM P31.20.2.2.2 and 4. It is possible that the ceiling might be finished around the pipe casing, when these items would not be required. It is assumed that the drain will be measured up to the top surface of the ground floor concrete slab.
	1	Cut circ. hole thro. q. tile pavg. for pipe 55 – 110 ϕ & m/gd. &	Enumerated and described in accordance with SMM P31.20.2.2.2-4. A hole may also be required in the kitchen worktop.
		Cut circ. hole thro. 25 swd. for pipe 55 – 110 ϕ.	
		Branch Discharge Pipe to W.C.	
	0.50	Pipes, st. nom. size: 90 ld. to BS 602 w. sold. jts & fxg. w. ld. tacks & brass scrs. plugd. to masonry @ 1 m ccs.	The branch discharge pipe for the WC is taken in lead to show the method of measuring connections of lead to the cast iron pipe and to the outlet of the vitreous china wc, respectively.

23.10

PLUMBING INSTALLATION (Contd.)

	1		E.O. ld. pipe, nom. size: 90, for made bend.	Enumerated extra over item as SMM Y10.2.1.
	1		E.O. ld. pipe, nom. size: 90, for spec. conn., brass soc. thimble w. sold. jt. to ld. pipe & c.m. (1:2) jt. to outlet of WC pan.	The thimble is enumerated and described in accordance with SMM Y10.2.2.1.1. uPVC (unplasticised PVC) pipes are jointed with a WC connector and solvent-welded joint.
			&	
			E.O. ld. pipe, nom. size: 90, for spec. conn., brass ferrule w.s.j. to ld. pipe & caulked ld. jt. to c.i. pipe.	Same procedure as for thimble.
	1		E.O. c.i. pipe, nom. size: 102, for fittg., 90° junctn. w. bolted c.i. inspectn. dr.	An enumerated extra over item in accordance with SMM Y10.2.4.6.1. The type of fitting is described as the pipe size > 65 mm.
	1		E.O. c.i. pipe, nom. size: 102, for fittg., 50 φ junctn.	Connection for bath and wash basin branch waste pipe, enumerated as a pipe fitting as SMM Y10.2.4.6.0.
			&	
			E.O. cop. pipe, nom. size: 38, for spec. jt. to c.i. pipe 50 φ w. 38 x 50 calkg. bush.	Enumerated item as SMM Y10.2.2.1.1.
			Pipe casgs. 1.560 2.400 3.960	Plastic or uPVC pipes to BS4514 can be jointed with ring seals or by solvent welding and may be supported by holderbats or clips.
2/	3.96		Timbr. casgs. wrot swd. 220 x 16 fxd. to swd. suppts. (m/s).	Pipe casings are measured on each face in accordance with SMM K20.1.2.1.0, and taken in metres giving a dimensioned description.
	1		E.O. ditto. for access panel, 450 x 150 in size, scrd. to swd. suppts. (m/s).	Enumerated extra over item as SMM K20.12.1.1. giving a dimensioned description.
	3.96		Butt jtd. suppts. 40 x 40 sn. swd., spaced @ 200 ccs. (crnr. longtdl. suppt.	Butt jointed supports are measured in metres where ≤ 300 mm wide, giving a dimensioned overall cross-section description and spacing of the members in accordance with SMM G20.11.2.1.0.

23.11

PLUMBING INSTALLATION (Contd.)

Timesing	Dimensions	Squaring	Description	Notes
2/	3.96		Butt jtd. suppts. 25x40 sn. swd., spaced at 200 ccs., p & s to masonry. (longtdl. suppts.	The method of fixing supports is to be stated where not at the discretion of the contractor (SMM G20.S2).
			3.960 less end suppts. 2/½/25 25 7) 3.935 562 220 less 65 155	
8/2/	0.16		Butt. jtd. suppts. 25x40 sn. swd. spaced at 560 ccs. (cross suppts.	The cross supports are at the wider spacings shown in the calculations.
2/	3.96 0.22		Paintg. gen. surfs., wd., k.p.s. & (3).	Painting to pipe casings measured in m^2 to general surfaces as the overall girth > 300 mm (SMM M60.1.0.1.0).
			Sundries	
	Item		Markg. the posn. of 11 nr. holes in the struct. for the cold water supply & foul drainage above grd. servs.	Item provided as SMM Y51.1.1.0.1. There are no chases and mortices in this case.
	Item		Allow for air testg. & commsng. the cold water supply & foul drainage above grd. servs.	A testing and commissioning item is required in accordance with SMM Y51.4.1.0.0.
				Note : hot water and heating installations are often covered by prime cost items, although the connections to appliances, the appliances and overflows are each measured.
			23.12	

14 Measurement of Drainage Work

ORDER OF TAKING OFF

A logical order of taking off and full annotation in waste is extremely important when measuring this class of work. Where separate drainage systems are to be provided for foul and surface water drainage, it is advisable to measure the drains and associated work in each system separately, although it is usually quicker and simpler to take all the manholes together.

A good order of measurement to adopt is outlined below.

(1) Lengths of main drain starting at the head of the drainage system and working downwards to the sewer or other point of disposal.
(2) Branch drains working in the same sequence.
(3) Connections, gullies and other accessories at the heads of the branch drains, linking up with the work previously measured under the drainage above ground installations. The building in of pipes at manholes are usually taken with the manholes.
(4) Manholes, measured in detail.
(5) Any other items, such as ventilating pipes, fresh air inlets, interceptors, connection to sewer, testing drains, and the like.
(6) Any septic tank installations, cesspits or soakways, measured in detail.

It is good policy to check that all rainwater, soil and waste or discharge pipes and gullies are shown connected to the drainage system on the drawings, since some lengths of branch drains are occasionally omitted.

DRAINS

The measurement of drains may often with advantage be broken down into three principal items.

(1) *Excavation of pipe trenches* is measured in metres, stating the commencing level where >0.25 m below existing ground level. The average depth range is given in stages of 250 mm (SMM R12.1.1.2.1). For example, where a drain trench runs from 1.10 m deep at one end to 2.70 m deep

at the other, the average depth is $\frac{1}{2} \times 3.80$ m = 1.90 m, assuming that the ground has a uniform fall over the length of the trench, and this will be classified as ≤2.00 m deep. Trenches to receive pipes ≤200 mm nominal size are grouped together and so described, whereas trenches for larger pipes are given separately for each nominal pipe size (SMM R12.1.1–2).

The pipe trench excavation item is deemed to include earthwork support, consolidation of trench bottoms, trimming excavations, filling with and compaction of general filling materials, and disposal of surplus excavated materials, and so these items do not have to be repeated in the trench excavated below groundwater level, next to roadways or existing buildings, or in unstable ground, this is to be incorporated in the item description. The or is unstable ground, this is to be incorporated in the item description. The method of determining the juxtaposition of the roadway or building to the pipe trench is illustrated in the *Code of Procedure for Measurement of Building Works*. Items are taken as extra over excavating trenches, irrespective of depth, for breaking out existing rock, concrete, reinforced concrete, brickwork, blockwork or stonework, or coated macadam or asphalt, measured in m^3, except for existing hard pavings which are measured in m^2 stating the thickness (SMM R12.2.1–2.1–5.0) The special requirements covering the excavation of pipe trenches next to existing live services or around existing live services crossing the trench should be noted (SMM R12.2.4–5.1.0).

(2) *Drain pipes* are described by the kind of pipe (such as clay, cast iron and pitch fibre), quality of pipe (such as British Standard and British Standard Surface Water clay), nominal size and method of jointing (SMM R12.8.1.1.0 and R12.S1 & 4). It is necessary to state that pipes are in trenches as distinct from in ducts, bracketed off walls or suspended from soffits, and these are each separately classified. Pipes are measured in metres over all fittings and branches (SMM R12.M7). Iron pipe runs ≤3.00 m in length are so described, stating the number, because of the relatively high cost involved. Vertical pipes, as in backdrops to manholes, are also kept separate. Pipe fittings, such as bends, junctions and diminishing pipes, are each described and enumerated as extra over the pipes in which they occur (SMM R12.9.1.1.0), with all cutting and jointing pipes to fittings deemed to be included (SMM R12.C4).

(3) *Concrete protection* may be provided to drains in three different forms which are described in SMM7 as beds; beds and haunchings; and beds and surrounds. These are each measured separately in metres, stating the width and thickness of beds and beds and haunchings, and the width, thickness of bed and thickness of surround in the case of beds and surrounds, and giving the nominal size of pipe in the latter two instances (SMM R12.4–6.1.1.0). Any formwork is deemed to be included in the concrete rates (SMM R12.C2). The illustrations in Fig. XIX show the various forms of concrete protection to drain pipes.

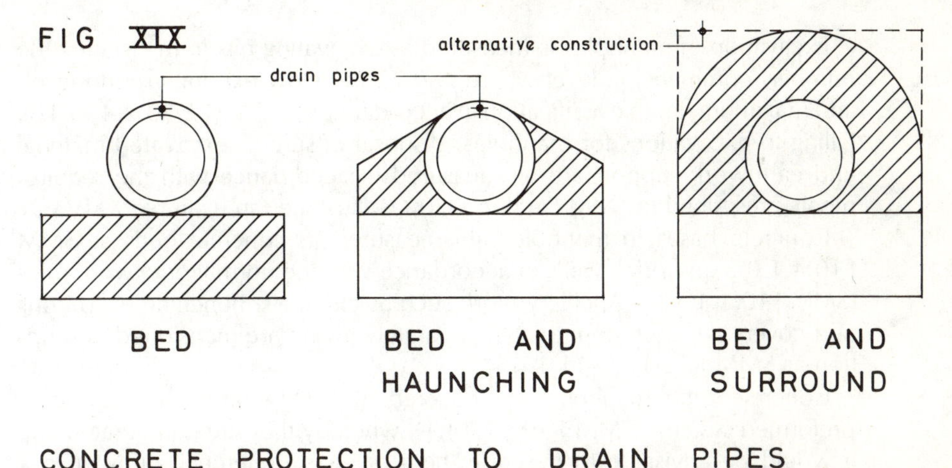

PIPE ACCESSORIES

Pipe accessories such as gullies, traps, inspection shoes, fresh air inlets, non-return flaps and the like are each enumerated and described (SMM R12.10.1.1.0). Dimensions stated for accessories are to include the nominal size of each inlet and outlet (SMM R12.D7). These items are deemed to include jointing pipes thereto and bedding in concrete, without the need for specific mention (SMM R12.C5). When connecting gullies to drains it is often necessary to insert one or two bends to obtain a satisfactory and smooth connection, and these are enumerated as extra over the pipes. Building in the ends of pipes at manholes is enumerated with a dimensioned description, including the provision of a brick arch where required, and the cutting of pipes is deemed to be included (SMM R12.C6).

MANHOLES

Manholes, as well as inspection chambers, septic tanks, cesspits and soakaways, are measured in detail under an appropriate heading, and it assists the estimator if the number involved is stated. A typical heading for manholes might read: 'The following in 8 nr brick manholes'.

Where a considerable number of manholes are encountered it is advisable to prepare a manhole schedule, detailing plan sizes, depths, wall thicknesses, connections, channel particulars, number of step irons, and details of cover slabs, covers, backdrops and any other special features. The taking off work is then considerably simplified and all the manholes are taken off together to avoid the repetition of similar items. A typical manhole schedule is illustrated in appendix IV.

Excavation of manholes is measured as excavating pits in m^3, stating the number, commencing level where > 0.25 m below existing ground level, and maximum depth classification in accordance with SMM D20.2.4.1–4.1. Filling to excavations for manholes, disposal of surplus excavated material and earthwork support are all measured in accordance with the requirements prescribed in SMM Section D20, as illustrated in example XXIV.

Concrete bases to manholes are measured as concrete beds as SMM E10.4.1.0.5 and brick walls in accordance with the normal brickwork rules (SMM F10.1.1.1.0). Ancillary work such as channels, benching, step irons and covers are each enumerated separately giving a dimensioned description as SMM R12.11.8–11.1.0.

Concrete tube manholes are covered in SMM7 under a heading of preformed systems (SMM R12.11.14.1), whereby they are enumerated and it would be advisable to describe them or give a reference number in a manhole schedule. Details of building in ends of pipes, channels, benching, step irons, covers and intercepting traps are to be stated.

Many manholes on domestic drainage systems could be more correctly described as inspection chambers, as they are not large enough to permit the entry of a man.

ASSOCIATED WORK

The work of connecting the drains to the public sewer is usually covered by a provisional sum in accordance with SMM A53.1.1.

This item may include all the drainage work outside the boundary of the site in the public highway, with a separate provisional sum for reinstatement of the highway, but sometimes the local authority is only prepared to make the actual connection, usually a saddle connection where an existing sewer is involved. Where the contractor has to carry out the drainage work in the public highway, the excavation work will have to be separately classified with separate items for breaking up and reinstatement of paved surfaces. SMM R12.16.1 provides for the enumeration and giving of details of the work in connecting to the local authority's sewer, where it is executed by the contractor (SMM R12.M10).

An item is taken for testing and commissioning the drainage system, stating the method of testing such as using water, smoke or air, and the number of stage tests (SMM R12.17.1.2.0). The contractor is sometimes required to test the manholes as well as the drains, and if this is the case it must be stated in the testing item.

WORKED EXAMPLE

A worked example follows covering the measurement of a house drainage system, including two brick manholes.

DRAINAGE WORK	Description	EXAMPLE XXIV
	MH1. – Sewer len. MHs 1-2 13.000 less 2/½/ int. len. of MHs 675 walls to MHs 2/215 430 1.105 11.895 depth MH.1 600 MH.2 975 2)1.575 av. depth 788 Tr. excavn.	The overall length of manholes are deducted when determining the length of pipe trench excavation, as the excavation for manholes is dealt with separately. The extra length due to the pipe being on the slope will cancel out the deduction needed for half the diameter of the sewer.
11.90	Exc. tr. for pipe ≤ 200 nom. size, av. depth of tr. ≤ 1.00 m. (MHs 1-2 len. MH2-sewer 4.000 less ½ MH (½/1.105) 553 3.447 depth MH2 975 sewer 1.200 2)2.175 av. depth to invt. 1.088 add. conc. bed 150 1.238	Trenches for pipes ≤ 200 mm nominal size are grouped together (SMM R12.1.1.2.0). The depth is given in average depth stages of 250 mm. The kind of pipe does not have to be specifically mentioned in pipe trench excavation descriptions.
3.45	Ditto. av. depth of tr. ≤ 1.25 m, in rdwy. (MH2-sewer	Work in roadways should be so described, because of the extra cost involved. The 500 mm length within the boundary of the site is not considered to be worth separating.
1.50 0.70	E.O. excvtg. trs. for breakg. out xtg. hd. pavgs., 50 th., conc. slabs & stackg for re-use.	Taking up slabbed paving is measured in m^2 as extra over pipe trench excavation, for breaking out existing hard pavings stating the thickness (SMM R12.2.2.2.0), with the additional disposal particulars.
1.50 0.70	E.O. ditto., coated macadam, 200 th.	A similar procedure is applied to the tarmacadam carriageway. Concrete bed is 700 mm wide
	Pipes len. of of pipe 11.895 add MH walls 2/215 430 12.325	The pipe lengths are greater than those of trench excavation, as the pipes pass through the manhole walls.
12.33	Pipes in trs., 100 pitch fibre to BS 2760 jtd. w. polypropylene couplgs. & rubber rgs. (MHs 1-2	The description of the pipes is to include the kind of material (pitch fibre), quality of pipe (to BS 2760), nominal size (100 mm) and method of jointing (polypropylene couplings and rubber rings) as SMM R12.8.1.1.0 and R12.S1 and S4.

24.1

DRAINAGE WORK Drawing No. 19

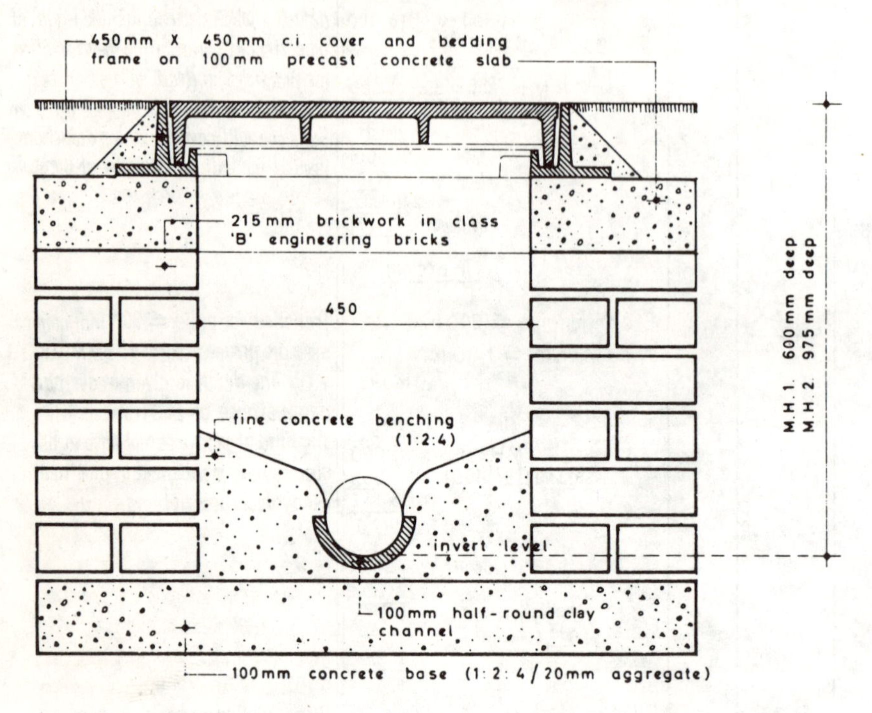

SECTION THROUGH MANHOLE scale 1:10

Note: All glazed vitreous clay pipes are surrounded with concrete, expansion joints in pipe surrounds at 10·0 m centres ; pitch fibre pipes have no concrete protection at all

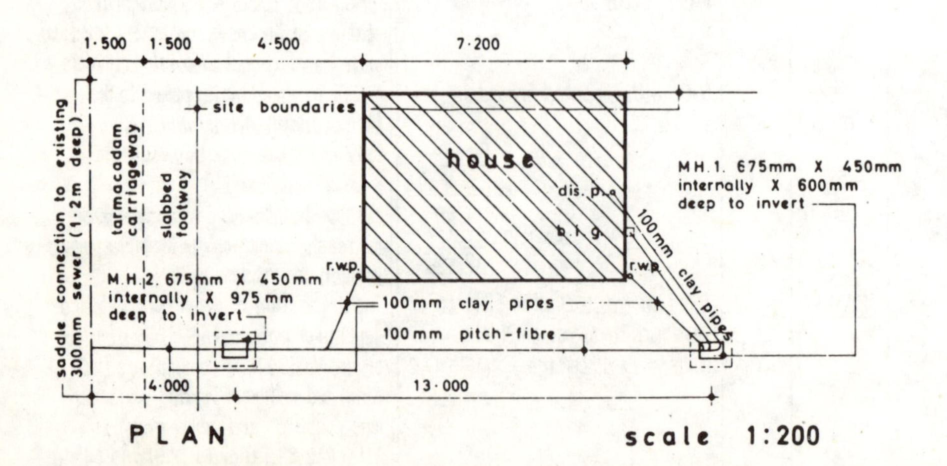

DRAINAGE WORK (Contd.)

Timesing	Dimension	Squaring	Description	Side notes
	1		E.O. 100 pitch fibre pipe for 100 x 100 junctn. w. coupler for clay pipe. len. of clay pipe 3·447 add MH wall 215 3·662 less ½ dia. of sewer 150 3·512	Junctions are enumerated as extra over the pipes in which they occur (SMM R12.9.1.1). The coupler needed in this case to connect with the clay pipe has been included in the description, as SMM R12.S5, although pipe fittings are normally deemed to include everything necessary for jointing (SMM R12.C4).
	3·51		Pipes in trs., 100 BS clay to BS 65 Pt. 2, jtd. w. flex. mech. jts. (sewer – MH.2 Conc. protectn. 3·587 less MH wall 215 3·372	Cement mortar joints to clay pipes have now been largely superseded by flexible mechanical joints, to permit slight movement of the pipes without fracturing the joints. See 'Building Technology' by the same author for typical examples.
	3·37		In situ conc. bed & surrd. (1:3:6/20 agg.), width of bed: 700 & thickness of bed & surrd. 150, nom. size of pipe: 100. (sewer – MH.2	Beds and surrounds to pipes are measured in metres, stating the width of the bed and the thickness of the bed and surround and the materials of which they are constructed, together with the nominal size of the pipe (SMM R12.6.1.1.0 and R12.S1). Any formwork is deemed to be included (SMM R12.C2).
			Branch drains head 300 MH1 600 2)900 450 add conc. bed 150 av. depth 600	Calculation of the average depth of branch pipe trenches, taking the pipes as 300 mm deep at the head of each branch.
	3·20		Exc. tr. for pipe ≤ 200 nom. size, av. depth of tr. ≤ 750. (b.i.g. to M.H.1	Excavation of branch pipe trenches in accordance with SMM R12.1.1.2.0. Where the disposal of surplus excavated material is to specified locations or backfilling is to be carried out using special materials, these must be included in the description.
	2·20		(rwp. to M.H.1	
	4·60		(disch. pipe to M.H.1 depth 300 830 2)1·130 565 add conc. bed 150 715	The lengths of excavation, pipes and concrete protection may vary because of the extensions of pipes through manhole walls and of vertical pipes. The lengths have been scaled from the drawings.
	2·10		Exc. tr. for pipe ≤ 200 nom. size, av. depth of tr. ≤ 750. (rwp to drain) 5·000 add vert. len. 300 5·300	

24.2

DRAINAGE WORK (Contd.)

Times	Dim.	Sq.	Description	Location	Notes
	3·50		Pipes in trs., 100 BS clay to BS 65 Pt. 2, jtd. w. flex. mech. jts.	(b.i.g. to M.H.1	If these were iron pipes in runs ≤ 3·00 m long, they would be so described, stating the number (SMM R12.8.1.1.1).
	5·30			(disch pipe to M.H.1	
	2·60			(rwp. to M.H.1	Note the extensive use of locational notes in waste.
	2·10			(rwp. to drain	
	3·20		In situ conc. bed & surrd., (1:3:6/20 agg.), width of bed: 700, thickness of bed & surrd: 150, nom. size of pipe: 100.	(b.i.g. to M.H.1	See SMM R12.6.1.1.0 for the measurement of beds and surrounds, giving the width and thickness of bed, thickness of surround and nominal size of pipe.
	2·20			(rwp. to M.H.1	
	4·60			(disch. pipe to M.H.1	
	2·10			(rwp to drain	Concrete can also be specified by strength, such as 11·50 N/mm².
			<u>Gullies & Connectns.</u>		
2/	1		Pipe accessory: gully, clay trapped to BS 539 w. 100 Q outlet, 75 back inlet & 150 x 150 sq. c.i. galvd. grtg.		Gullies are enumerated with a dimensioned description as SMM R12.10.1.1.0. Jointing to pipes and bedding in concrete are deemed to be included and so do not require specific mention (SMM R12.C5). Dimensions stated for accessories are to include the nominal size of each inlet and outlet (SMM R12.D7).
	1		Pipe accessory: gully, clay trapped to BS 539 w. 100 Q outlet, 50 back inlet & 150 x 150 sq. c.i. galvd. grtg.		Back inlet gullies are also required at the base of rainwater pipes as they discharge into a combined drain.
3/	2		E.O. 100 clay pipe for bend.	(gullies	Bends are enumerated as extra over pipes (SMM R12.9.1.1), and cutting and jointing pipes are deemed to be included (SMM R12.C4). Two bends are taken to each square gully (one for direction and the other for gradient).
	1			(disch. pipe	
			<u>Manholes</u>		Manholes should be given under a suitable heading, preferably stating the number of manholes (SMM R12.11).
			<u>The follg. in 2 nr. bk. manholes</u>		

24.3

DRAINAGE WORK (Contd.)

			len. width 675 450 add walls 2/215 430 430 1·105 880 depth MH.1 MH.2 to invert 600 975 add conc. base, chan. & beddg. 150 150 750 1·125	These manholes would be more accurately described as inspection chambers, as SMM R12.12, as they are not large enough to permit a man to enter them. Note: half brick walls are often considered suitable for manholes ≤ 900 mm deep, when built in engineering bricks. Schedules should be used when more than a few manholes are involved (See Appendix IV).
	1·11 0·88 1·13		Exc. pit, max. depth ≤ 2·00 m. (In 1 nr.) (MH.2	Excavation for manholes is measured in m^3 as pits, adopting the depth classification given in SMM D20.2.4.1–4.0 and stating the number.
	1·11 0·88 0·75		Ditto. max. depth ≤ 1·00 m. (In 1 nr.) (MH.1	
	1·11 0·88 1·13		Disposal of excvtd. mat. off site. (MH.2	All excavated soil is to be removed from the site, as no backfill is required (See SMM D20.8.3.1.0).
	1·11 0·88 0·75		(MH.1	
			Earthwk. suppt. 1·105 880 2/ 1·985 3·970	
	3·97 1·13		Earthwk. suppt. max. depth ≤ 2·00 m, distance between opposg. faces ≤ 2·00 m. (MH.2	Earthwork support is measured in m^2, stating the maximum depth and distance between opposing face ranges as SMM D20.7.1–3.1–3.0.
	3·97 0·75		Ditto., max. depth ≤ 1·00 m, do. (MH.1	
2/	1·11 0·88		Compactg. bott of excavn.	Compacting the bottom of excavation is measured in m^2 (SMM D20.13.2.3.0).
2/	1·11 0·88 0·10		In situ conc. bed (1:2:4/20 agg.), thickness ≤ 150, poured on or against earth.	Beds of concrete are measured in m^3 in accordance with SMM E10.4.1.0.5, and giving the appropriate thickness and placing classifications.

24.4

DRAINAGE WORK (Contd.)

len. of bk. walls.

	675
	450
2/	1·125
	2·250
add crnrs 2/215	860
	3·110

depths

	MH.1	MH.2
depth to invt.	600	975
less conc. slab & m.h. cover	200	200
	400	775
add chan. & bed	50	50
	450	825

Note the method of building up wall dimensions in waste.

Timesing	Dimensions	Description	Notes
	3·11 0·45 3·11 0·83	Bk. wall, facewk. o.s., thickness: 215, class B eng. bks. to BS 3921 in English bond in c.m. (1:3) & ptg. w. flush jt. as wk. proceeds. (MH.1 / (MH.2	The brick walls are measured in m², stating the thickness of wall, facework one side (SMM F10.1.2.1.0), with the kind, quality and size of bricks, type of bond, composition and mix of mortar, and type of pointing as SMM F10.S1–4. Brick sizes are normally given in a preamble clause. Facework is any work in bricks or blocks finished fair (SMM F10.D2).
2/	1	In situ conc. benchg. (1:2:4/10 agg) to bott of MH, 675 × 450 & av. 180 th., floated w.c.m. (1:2) screeded smth. fin. to falls to chans.	Benching is enumerated with a dimensioned description (SMM R12.11.9.1.0).
	1	Chan. 100 dia. h.r. clay st. 675 lg. & beddg. on conc. benchg. (m/s) & jtg. in c.m. (1:2). (MH.2	The channels are enumerated with a dimensioned description as SMM R12.11.8.1.0.
	1	Chan. 100 dia. h.r. clay curvd. 1·10 m gth., do. (MH.1	No step irons will be required to these shallow manholes; where needed they are enumerated giving a dimensioned description (SMM R12.11.10.1.0).
3/	1	Chan. 100 dia. 3/4 sec. bend, 250 gth. & beddg. & jtg. a.b. (MH.1	
	4 2	B.i. end of 100 pipe into bk. wall, thickness: 215 & m/gd. (MH.1 / (MH.2	Enumerated item as SMM R12.11.7.1.0. **Building in ends of pipes** is deemed to include cutting pipes (SMM R12.C6).

24.5

DRAINAGE WORK (Contd.)

2/	1		Precast conc. cover slab (1:2:4/20 agg.) 1105 x 880 x 100 th., reinfd. w. stl. fabric to BS 4483, ref. A193, weighg. 3·02 kg/m², w. opg. to rec. 450 x 450 cover (m/s), fin. w. smth. upper surf. & settg. in c.m. (1:3).	The precast concrete cover slab is enumerated with a dimensioned description, including reinforcement details, as SMM F31.1.1.0.1. The description should also include the materials and mix, bedding and fixing and surface finishes (SMM F31.S1–5).
2/	1		Cover & fr., galvd. c.i. to BS497 Pt. 1, grade C, single seal flat type, 457 x 457 in size, weighg. 31 kg. & settg. fr. in c.m. (1:3) & beddg. cover in grease.	Covers are enumerated with a dimensioned description as SMM R12.11.11.1.0. If the covers were required to carry vehicular traffic, a heavier variety would be needed.
			End of manholes	
			Provide the Provsnl. Sum of £60 for saddle conn. of drain to sewer to be carried out by L.A.	A provisional sum for the sewer connection as SMM A53.1.1. and R12.M10.
			Provide the Provsnl. Sum of £80 for perm. reinstatement of the rdwy. & footway to be carried out by the H.A.	A provisional sum is also inserted for the permanent reinstatement work by the highway authority as SMM A53.1.1.
	Item		Allow for testg. & commisng. drainage installtn. after backfillg. w. water test of not less than 1·50 m head.	Item stating the method of testing as SMM R12.17.1. If stage tests are required, these are to be listed. Provision of water and other supplies and test certificates is deemed to be included (SMM R12.C7–8).
	Item		Allow for disposal of surf. water.	Item for keeping excavations free of surface water as SMM R12.3.1. Where excavation is measured below groundwater level, an item is included for disposal of groundwater (SMM R12.3.2).

24.6

15 Measurement of External Works

This chapter is concerned with the measurement of roads, drives, paths, grassed areas, planting of trees and shrubs and fencing. Drainage work has already been covered in the previous chapter. Demolition work is considered to be outside the scope of this book and has accordingly been omitted, but is included in *Advanced Building Measurement*.

ROADS, DRIVES AND PATHS

The major components of roadworks are covered in work sections Q20–25, embracing the various types of pavings and their associated sub-bases; with kerbs, edgings and channels incorporated in work section Q10.

In general, the sub-base or foundation for the road or footpath and the surfacing are each measured separately with a full description. Filling material is measured in m^3, classifying the average thickness as $\leq$ or $>$0.25 m, while the approach to the measurement of the surfacing will vary according to the nature of the material. For example, *in situ* concrete is measured in m^3 with a thickness classification as SMM E10.4.1–3.0.1, while coated macadam and asphalt are measured in m^2, stating the thickness and number of coats (SMM Q22.1.1.2.0), and paving slabs in m^2, stating the thickness (SMM Q25.2.1.2.1). Compaction of filling or bottom of excavation is taken as a superficial item (SMM D20.13.2.2–3.0). Certain incidental items are measured separately, such as forming expansion joints in concrete roads (SMM E40.2.1–2.1.0), and trowelling concrete in channels and around gullies (SMM E41.3.0.0.2).

With concrete roads, the waterproof membrane and steel reinforcement are also measured separately (the waterproof membrane and fabric reinforcement in m^2 and bar reinforcement in tonnes). Formwork to the edges of concrete road slabs, except at expansion joints, is measured in metres and classified in the appropriate height category (SMM E20.2.1.2–4.0).

When measuring footpaths, consisting of paving slabs, it is necessary to state the kind, quality, size, shape and thickness of slabs, nature of surface finish, method of bedding, treatment and layout of joints and nature of base (SMM Q25.S1–7). This type of information is often incorporated in preamble clauses or may be the subject of cross-references to the project specification where one has been prepared adopting the co-ordinated project information approach, with the same component references in drawings, specification and billed descriptions.

Precast concrete kerbs, channels and path edging are each measured separately in metres with a dimensioned description and often giving a catalogue or other reference number, and including the foundation and haunching (SMM Q10.2–4.1.0.2). Curved members are so described stating the radii.

Any surface water drainage work to roads would be measured in the manner outlined in chapter 14.

GRASSED AREAS

Filling to make up levels with topsoil is measured in m^3 and stating the average thickness as to whether $\leq$ or >0.25 m and other relevant particulars as SMM D20.10.1–2.1–3.3. Cultivating, surface applications, seeding and turfing are each measured separately in m^2, stating the depth of cultivating, the type and rate of surface applications and rate of seeding. Full descriptions accompany these items, paying particular attention to such matters as kind, quality, composition and mix of materials, method of application, method of cultivating and degree of tilth, timing of operations and method of securing turves (SMM Q30.S1–5). The type of surface applications must be stated and they include herbicides, selective weedkillers, peat, manure, compost, mulch, fertiliser, soil ameliorants and sand (SMM Q30.D1). Cultivating is deemed to include the removal of stones and seeding includes raking or harrowing in and rolling (SMM Q30.C1&3).

TREES, SHRUBS AND HEDGES

Trees are enumerated, giving the botanical name, BS size designation and root system or, alternatively, the girth, height, clear stem and root system, and the description can include supports and ties, and refilling with special materials and watering where required (SMM Q31.3.1.1–2.3–5). Shrubs are also enumerated stating the botanical name, height and root system (SMM Q31.5.1.1.4–5). Where trees are to be provided with tree guards, these are enumerated with a dimensioned description (SMM Q31.10.1.1).

Hedge plants are usually measured in metres, giving the botanical name, height, spacing, number of rows and layout (SMM Q31.6.1.2), as illustrated in example XXV.

FENCING

Fencing is measured in metres, stating the type and height of fencing (measured from the surface of the ground or other stated base to the top of the infilling or the top wire or rail where there is no infilling), and the spacing, height and depth of supports (SMM Q40.1.1.1.0). The description of the fencing shall include the kind and quality of materials, construction, surface treatments applied before delivery to site and size and nature of backfilling.

Fencing posts and struts occurring at regular intervals, classified as supports, are included in the description of the fencing, without the need for separate measurement (SMM Q40.D1). Special supports such as end posts, angle posts, integral gate posts and straining posts, with supporting struts or back stays, are enumerated separately as special supports as extra over the fencing in which they occur (SMM Q40.2.1–4.1.1–2). The excavation of holes for supports, special supports and independent gate posts, together with backfilling, disposal of surplus materials and earthwork support are deemed to be included (SMM Q40.C1), while concrete surrounds to posts and stays are included in the descriptions of special supports and independent gate post descriptions. Gates are enumerated, stating the type, height and width, and are deemed to include gate stops, gate catches, independent gate stays and their associated works (SMM Q40.5.1.1.0 and Q40.C5).

It should be noted that fencing to ground sloping $> 15°$ from the horizontal, fencing set out to a curve but straight between posts, curved fencing radius > 100m, curved fencing radius ≤ 100 m stating the radius, and lengths ≤ 3 m are each measured separately.

The measurement of brick and stone boundary walls was covered in chapters 5 and 6.

WORKED EXAMPLE

A worked example follows covering the measurement of a road, channels, footpath, edging, fencing, gates, trees, shrubs, hedges and a grassed area.

EXTERNAL WORKS

EXAMPLE XXV

Access Road

		len.
		24.000
add bellmth.		3.000
		27.000

	bellmth.
	3.000
less chan.	254
	2.746

Note use of sub-headings which act as signposts throughout the dimensions.

Timesing	Dimensions	Description
	27.00 3.00	Exc. topsoil for preservtn. av. 225 dp. &
	15.00 4.50	Compactg. bott. of excavn.
2/3/14/	2.75 2.75	(bellmth.

Excavating topsoil which is required to be preserved is measured in m², stating the average depth (SMM D20.2.1.1.0).
It is assumed that the topsoil on the site is 225 mm deep, otherwise part of the excavation would have to be taken as excavation to reduce levels in m³ (See SMM D20.2.2.1.0).
The additional area at each side of the bellmouth = $\frac{3}{14}$ x radius².

Compacting bottom of excavation is measured in m² as SMM D20.13.2.3.0.

Timesing	Dimensions	Description
2¼/	27.00 3.00 0.10	Fillg. to make up levs., av. thickness ≤ 0.25 m, arisg. from excavns., topsoil.
2¼/	15.00 4.50 0.10	
2¼/2/3/14/	2.75 2.75 0.10	

		len.
		15.000
less chan. 2/254		508
		14.492

		width
	4.500	3.000
less chans. 2/254	508	508
	3.992	2.492

The spreading of the topsoil is assumed to be carried out direct from the access road excavation without the need for on site spoil heaps, and is thus measured in m³ in accordance with SMM D20.10.1.1.3, stating the appropriate thickness classification. The timesing factor of 2¼ takes account of the variations in depth of the excavated and deposited topsoil (225 and 100 mm respectively).
In practice the plan would normally show both ground and finished levels from which the quantities of fill would be computed, as described in Chapter 3.

Timesing	Dimensions	Description
	27.00 2.49 0.15	Fillg. to make up levs., av. thickness ≤ 0.25 m, hardcore consistg. of gravel rejects obtnd. off site.
	14.49 3.99 0.15	
2/3/14/	3.00 3.00 0.15	

The hardcore filling in making up levels as road sub-base is measured in m³ in accordance with SMM D20.10.1.3.0.

25.1

EXTERNAL WORKS Drawing No. 20

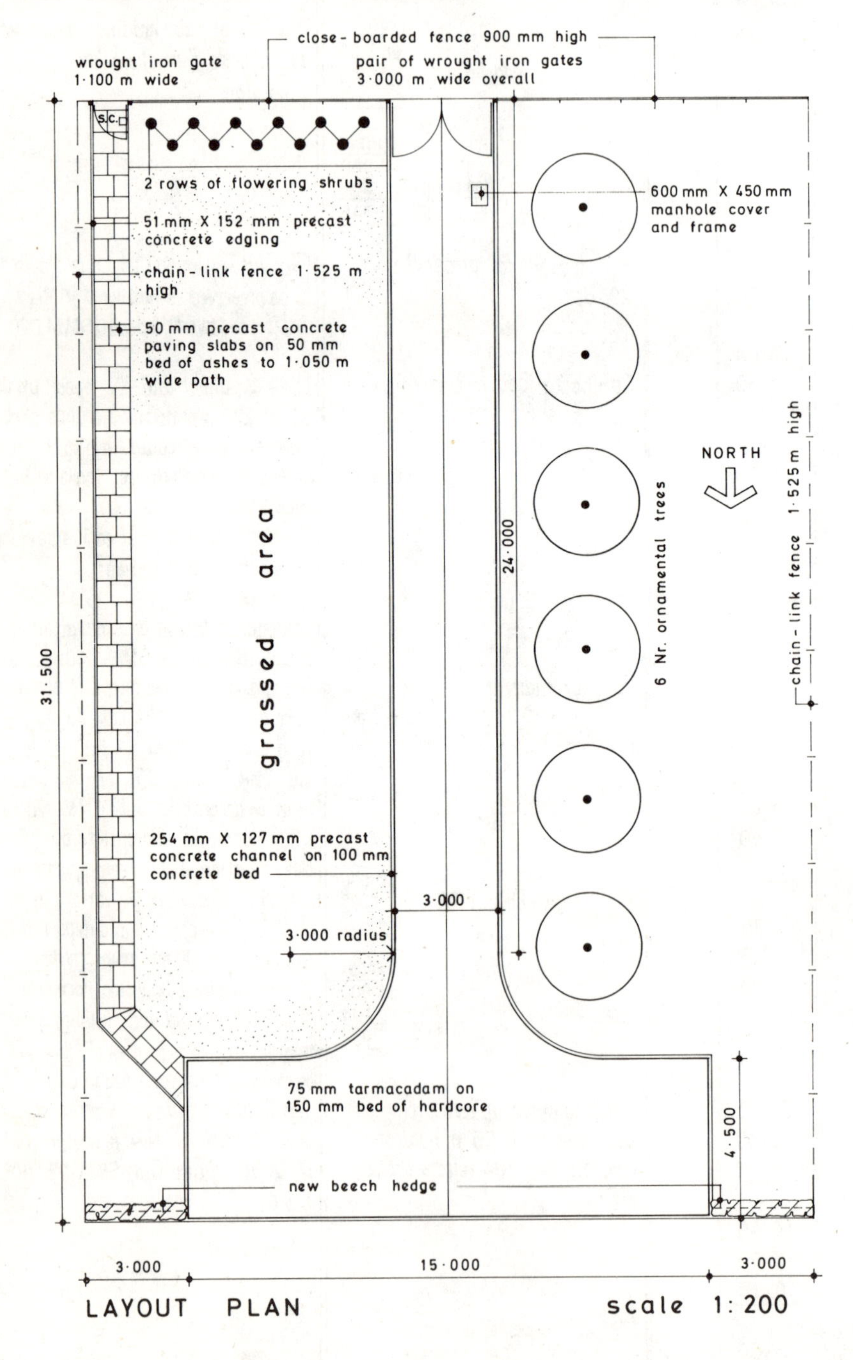

LAYOUT PLAN scale 1: 200

EXTERNAL WORKS (Contd.)

	27.00 2.49		Compactg. fillg., blindg. w. hoggin.
			&
	14.49 3.99		Road tarmcdm. to BS 802 w. slag. agg. in 2nr. cts. consistg. of base course 55th. to table 1a of 38 nom. size mat. & med. textured wearg. course, 20th. laid warm to table 3 of 13 nom. size mat. laid to falls & crossfalls & to slopes ≤ 15° from horizontal, & consolidated w. 8 tonne roller on blinded hdcore base (m/s).
2/3/14/	3.00 3.00		

	Chans.
	15.000
less access rd. 2.492 rad. 2/3.000 6.000	8.492
	6.508

	15.00		Precast conc. chan. 254 x 127 w. gravel agg. to BS 340, fig. 8, b.& j. in c.m. (1:2), on in situ conc. fdn. (1:3:6/20 agg.), 254 x 100.
	6.51		
2/	3.99		
2/	24.00		
2/¼/2/22/7/	3.00		Precast conc. chan 254 x 127, do., curved to 3.00 m ext. rad., on in situ conc. fdn. a.b.

	Path width
	1.050
add edging and conc. haunchg.	100
	1.150

25.2

The compacting of filling item includes the blinding, stating the material used (SMM D20.13.2.2.1).
The tarmacadam road is measured in m², giving the particulars listed in SMM Q22.S1–6, with the details as to thickness, number of coats and falls listed in SMM Q22.1.1.2.0.
Alternatively the reference could be to BS 4987: coated macadam for roads and other paved areas.
The making good and labour finishing the surface of the tarmacadam edge of the manhole cover frame is deemed to be included (SMM Q22.C1b).

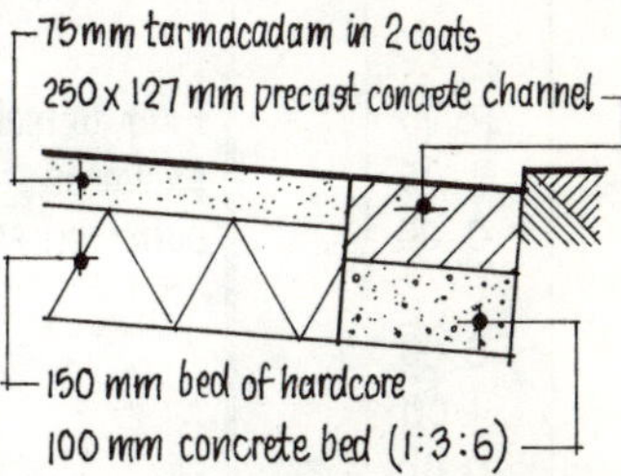

Precast concrete channels are measured in metres with a dimensioned description and including the concrete foundation (SMM Q10.4.1.0.2). Foundations are deemed to include formwork (SMM Q10.C2).
The purpose of the channel is to strengthen and define the edges of the road.
Curved work is kept separate stating the radius.
Allowance is made for the edging and backing when measuring the width of excavation of topsoil. The lengths are scaled from the drawing.

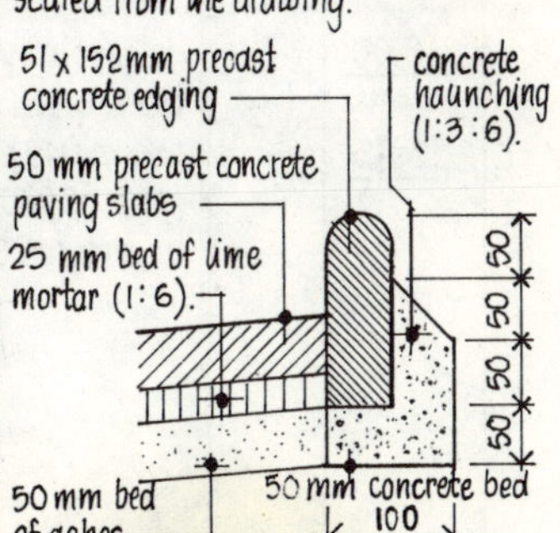

EXTERNAL WORKS (Contd.)

Dimensions	Description	Notes
25.75 1.15 3.00 1.15	Exc. topsoil for preservtn. av 125dp. & Compactg. bott of excavn.	Excavation of topsoil and compacting bottom of excavation is measured in m^2, while filling to make up levels with topsoil is measured in m^3
25.75 1.15 0.13 3.00 1.15 0.13	Fillg. to make up levs., thickness ≤ 0.25m, arisg. from excvns. topsoil.	
25.75 1.05 0.05 3.00 1.05 0.05	Fillg. to make up levs., av. thickness ≤ 0.25m, ashes obtnd. off site.	Ash beds are measured in m^3, stating that the average thickness is ≤ 0.25m as SMM D20.10.1.3.0, followed by a superficial compacting item.
25.75 1.05 3.00 1.05	Compactg. fillg. & Precast conc. paving slabs 50th.w. gravel agg. to BS 368, 600 x 600 & 450 x 600 in size laid bkg. jt. w 150 lap & grtd. in l.m. (1:6) on wet screeded bed of l.m. (1:6), 25 th. on ashes (m/s), ld. to falls, crossfalls & to slopes ≤ 15° from hor.	Paving slabs or flags are measured in m^2 giving the particulars listed in SMM Q25.2.1.2.1. and Q25. S1-8. These include the thickness, slope, kind and quality of materials, size and shape of units, bedding, treatment and layout of joints and nature of base. All fair joints, cutting and working around obstructions is deemed to be included (SMM Q25.C1).
	<u>Edging</u>	
26.00 3.75	Precast conc. edgg. 51 x 152 w. rdd. top & gravel agg. to BS 340, b.&j. in c.m (1:2), on in situ conc. (1:3:6/20 agg.) fdn. 100 x 50 & haunchg. 50 x 70 av. depth.	Edging is measured in metres with a dimensioned description, including foundation and haunching as SMM Q10.3.1.0.2. Edgings are deemed to include cut angles and ends, and foundations and haunching include formwork (SMM Q10.C1-2). Excavation has already been taken with the path.

EXTERNAL WORKS (Contd.)

			Description	Notes
			Grassed area len. 24.000 add curved end 2.746 26.746 less shrub border 1.800 24.946	Calculation of lengths of grassed area in waste. The width has been scaled from the drawing.
	24.95 7.30		Cultivatg. soil 100 dp. by hand or ped. optd. mach., evenly graded to fin. levs. to a fine tilth. & Fertilisg. cultvtd. soil (m/s) w. appvd. grassld. fertiliser basic price 60p/kg, at a rate of 70g/m². & Seedg. cultvtd. soil (m/s) in 2 nr. operatns. w. appvd. grass seed, basic price £2.50/kg at the rate of 70g/m².	Cultivating, surface applications, seeding and turfing are each measured separately in m² (SMM Q30.1–4). The cultivating description is to include the depth, method of cultivating and degree of tilth. (SMM Q30.1.1.0.0. and Q30.S2). Fertilising (one of the surface applications listed in SMM Q30.D1) is to state the type of fertiliser and rate of application, and is deemed to include working in (SMM Q30.2.1.0.0 and Q30.C2). The seeding description is to include the rate of application, kind and quality of material and is deemed to include raking or harrowing in and rolling (SMM Q30.3.1.0.0. and Q30.C3). Note also the inclusion of basic prices for these materials. Details regarding the timing of the operations is likely to be included in a preamble clause or a project specification.
3/14/	2.75 2.75		Ddt last three items. (bellmth.	Adjustment of grassed areas for the two irregular corners.
½/	1.50 1.50		(spld. corner	
			Shrubs	
11/	1		Shrubs, Berberis darwinii, ht. 450, w. balled fibrous rt. system, inc. excvtg. pit 450 x 450 x 450 dp. & refillg. w. excvtd. topsoil mixed w. 20% compost, remove surplus excvtd. mat. & waterg. as specfd.	Shrubs are enumerated giving the botanical name of the shrub and describing the height and root system, size of pit to be excavated, material used for refilling, removal of surplus excavated material and any watering requirements as SMM Q31.5.1.1.4–5.

25.4

EXTERNAL WORKS (Contd.)

Timesing	Dimension	Squaring	Description	Commentary
			Trees	
6/	1		Malus tschonoski, lt. stand., gth. 40–50, ht. 1·80–2·10 m, clear stem 1·20–1·50 m, w. adequate compact rt. system, supptd. by hdwd. stake, 2·00 m x 50 dia. & plastic tree tie, excavtg. pit 900 x 900 x 600 dp. & refillg. w. excvtd. topsoil incorptg. 20% manure, remove surplus excvtd. mat. & waterg. as specfd.	A similar procedure is adopted for trees, but including the girth measured 1 m above ground and height of clear stem. Alternatively, the BS size designation can be stated. This item may also include the provision and fixing of a tree stake and tree tie (SMM Q31.3.1.1–2.3–5).
6/	1		Tree guards, plastic tubes, perforated, 1·20 m lg. x 150 dia.	Tree guards are enumerated with a dimensioned description (SMM Q31.10.1.1).
			Hedges	
2/	3·00		Hedge plants, beech, ht. 450, spaced 375 apart in 2 rows in staggered formatn. inc. excavtn., refillg. w. topsoil & fertiliser & waterg. as specfd.	Hedge plants can be enumerated or measured in metres. When measured in metres, the description is to include the height, spacing, number of rows and layout (SMM Q31.6.1.2.4–5). Alternatively reference could be made to the project specification clauses for detailed requirements.
			Chainlk. fences	
2/	31·50		Fencg. chnlk. to BS 1722, Pt. 1, 1·525 m hi., w. 3 dia. green plastic coated galvd. m.s. fabric of 51 mesh, tied to 3 nr. galvd. m.s. line wires of 3·25 mm dia. w. galvd. m.s. tying wire @ 450 ccs. & straing. & threadg. line wires thro' & inc. coated 45·5 x 45·5 x 4·67 mm m.s. L inter posts 2·160 m lg. spaced @ 3·00 m ccs. & driven 610 into the grd., leavg. 1·55 m above g.l.	Fences are measured in metres, stating the type, height of fence, and spacing, height and depth of supports (SMM Q40.1.1.0). Supports are posts, struts and the like occurring at regular intervals. The height of fencing is measured from the surface of the ground or other stated base to the top of the infilling (SMM Q40.D3), and the steel posts extend 25 mm above the top of the infilling. Special supports such as end posts, angle posts and straining posts, and their supporting struts, are enumerated as extra over the fencing in which they occur, stating the size, height and depth (SMM Q40.2.1–5.1.2). Details are required of the kind and quality of materials, construction, surface treatment before delivery, and size and nature of backfilling (SMM Q40.S1–4).
2/	2		E.o. chnlk. fencg. for spec. suppt., coated 63·5 x 63·5 x 6·22 mm m.s. L strng. post 2·16 m lg. let 610 into the grd. & 1 nr. coated 45·5 x 45·5 x 4·67 mm m.s L strut. 2·16 m lg., bolted to strng post, inc. surrdg. botts. of posts & struts w. conc. (1:12/40 agg.) 450 x 450 x 450 dp.	

25.5

EXTERNAL WORKS (Contd.)

		Description	Notes
			The length of fencing descriptions could be reduced substantially by referring to the appropriate project specification clauses. Fencing work is deemed to include excavating holes for supports, special supports and independent gate posts, backfilling and disposal of surplus materials and earthwork support (SMM Q40.C1). This only leaves the concrete to the special support holes to be described in the enumerated items. Fencing to ground sloping > 15° from the horizontal, set out to a curve but straight between posts, curved to radius > 100 m, and curved to radius ≤ 100 m stating the radius, are each so described.
		Close bdd. fence	
	0.35 7.30 9.00	Fencg. clse. bdd. English oak to BS 1722 Pt.5, 900 hi. pressure creosoted, consistg. of 2 nr. ex. 75 x 75 sawn arris rails hsd. to posts, w. 89 x 19 sawn feather edged pales w. 13 overlappg. jts., ea. pale 2ce nailed to rls. ex. 63 x 38 2ce wethd. wrot. cappg., 25 x 150 sawn gravel bd. nailed to & inc. 150 x 63 x 25 sawn cleats nailed to posts, 100 x 100 x 1.50 m lg. sawn posts. w. ptd. bases & wethd. tops, spaced @ 2.75 m ccs. driven 600 into the grd., leavg. 900 above g.l.	Close boarded fences are measured in metres with a full description as SMM Q40.1.1.1 and Q40.S1–4. The posts are included in the description of the fencing, stating the size, height above surface of ground and depth below surface, and pointing and driving posts into the ground.
2/	2	E.o. clse. bdd. fence for pressure creosoted English oak, integral gate posts, 200 x 200 x 1.50 m lg. & let 600 into the grd., inc. surrdg. botts of posts w. conc. (1:12/40 agg.) 600 x 600 x 450 dp.	The integral gate posts are enumerated as extra over the fencing in which they occur as SMM Q40.2.3.1.0.
		Gates	Gates are enumerated, stating the type, height, width and method of construction. Gates are deemed to include gate stops, gate catches and independent gate stays and their associated works (SMM Q40.C5), and so these items and their fixing do not require inclusion in the gate description.
	1	Gates (1 nr. pr.) w.i. coated ornamental to BS 4092 Pt.1, size 3.00 m x 900 h/a to cat. ref. 'X' of manufacturer 'A', inc. all assctd. fittgs.	

25.6

EXTERNAL WORKS (Contd.)

Timesing	Dimensions	Squaring	Description	Side notes
	1		Gate w.i. coated ornamental to BS 4092 Pt. 1, size 1·10 m x 900 o/a to cat. ref. 'Z' of manufacturer 'C', inc. assctd. fittgs.	In these examples reference is made to a manufacturer's catalogue in accordance with SMM General Rules 6.1. As an alternative approach a basic price could be given to assist the estimator in pricing each item.
2/	3·00 0·90		Paintg. gates, met., ext., ornamental type, prep. & ③.	Painting to gates is measured overall regardless of voids on each side in m^2 as SMM M60.7.3.0.0. and M60.M12. No priming is included in the description as the British Standard requires protection treatment of the gates to be carried out at the manufacturer's works.
2/	1·10 0·90			

16 Bill Preparation Processes

This chapter starts with an examination of the traditional method of bill preparation and then leads on to an investigation of the more recent processing techniques with their merits and demerits.

WORKING UP

A description is given of the final stages leading up to the preparation of bills of quantities for building work by the traditional method after the dimensions have been taken off in the manner described and illustrated in the earlier chapters. The term 'working up' is applied to all the subsequent operations collectively and consists of the following processes.

(1) Squaring the dimensions and entering the resultant lengths, areas and volumes in the third or squaring column on the dimensions paper.
(2) Transferring the squared dimensions to the abstract (illustrated in Example XXVI), where they are written in a recognised order, ready for billing, under the appropriate work section headings, and are subsequently totalled and reduced to the recognised units of billing, preparatory to transfer to the bill.
(3) In the bill of quantities, part of one being illustrated in Example XXVII, the various items of work which together make up the complete building are listed under the appropriate work section headings, with descriptions printed in full and quantities given in the recognised units of measurement, as laid down in the *Standard Method*. The bill also contains rate and price columns for pricing by contractors when tendering for the project.

BILLING DIRECT

The traditional working up process which has been used for many decades in quantity surveyors' offices is very lengthy and tedious, and various ways of shortening the process have been developed. One of the first methods to be introduced was to 'bill direct', by transferring the items direct from the dimension sheet to the bill, thus eliminating the need for an abstract, and so saving both time and money.

The billing direct system is most suitable where the number of like items is limited and the work is not too complex in character. Drainage work is a particular instance where this shorter method can, with advantage, be adopted on occasions.

With the object of speeding up the working up process and reducing the labour involved, further methods using computers, sometimes on a national basis, or a 'cut and shuffle' system in the quantity surveying office have been developed. These newer methods will be described later in this chapter.

SQUARING DIMENSIONS

The term 'squaring dimensions' refers to the calculation of the numbers, lengths, areas and volumes and their entry in the third or timesing column on the dimensions paper. The following examples illustrate the method of squaring typical dimensions on dimensions paper.

	Dimensions			*Notes*
12/	4.50	54.00	Flr. membrs., 50 × 175 sn. swd.	Linear item: Total length is 54 metres, and no further reduction is required in the abstract.
	5.78 5.48	31.67	Exc. topsoil for preservn. av. 150 dp.	Square or superficial item: area is 31.67 m² which will be reduced to 32 m² in the abstract, as the fraction exceeds one-half.
	19.50 0.75 0.23	3.36	*In situ* conc. fdns. (1:3:6/40 agg.) poured on or against earth.	Cubic item: Volume of concrete is 11.46 m³ which will be reduced to 11 m³ in the abstract, as the fraction is less than one-half. Note method of casting up a number of dimensions relating to the same item with the total quantity entered in the squaring column.
	30.00 0.90 0.30	8.10		
		11.46		Deductions following the main items can be dealt with in a similar manner, as shown in example I.

Steelwork will often be transferred to the abstract as enumerated items where they will subsequently be reduced to tonnes.

When there are timesing figures entered against the item to be squared, it is often simpler to multiply one of the figures in the dimension column by the timesing figure before proceeding with the remainder of the calculation. Alternatively, the total obtained by the multiplication of the figures in the dimension column is multiplied by the timesing figure.

The squaring must be checked by another person to eliminate any possibility of errors occurring. All squared dimensions and waste calculations should be ticked in coloured ink or pencil on checking and any alterations made in a similar manner. Amended figures need a further check. Where, as is frequently the case, calculating machines are used for squaring purposes a check should still be made.

ABSTRACTING

Transfer of Dimensions

Typical completed abstract sheets are produced in Example XXVI, later in this chapter, and the items will subsequently be produced in bill form in Example XXVII. The abstract in Example XXVI incorporates the dimensions of the foundations to a small building which formed Example I in chapter 4, where the dimensions were squared in readiness for transfer to the abstract sheets. As each item is transferred to the abstract, the description of the appropriate dimension item is crossed through with a vertical line on the dimension sheet, with short horizontal lines each end, so that there shall be no doubt as to what has been transferred.

The following example illustrates this procedure.

	Dimension	Squared	Description
	5.78 5.48	31.67	Exc. topsoil for preservn. av.150

Subdivisions of Abstract

The abstract sheets are ruled with a series of vertical lines spaced about 25 mm apart and are usually of A3 width.

Each abstract sheet is headed with the project reference, sheet number and work section, and possibly the sub-section of the work to which the abstracted dimensions refer. The majority of bills and abstracts are subdivided into work sections in the manner adopted in the *Standard Method*.

Each work section is usually broken down into a number of subsections as appropriate on the lines indicated in the following examples, and generally adopting the subdivisions contained in the *Standard Method*.

D20 *Excavating and filling:* site preparation; excavating; earthwork support; disposal of water; disposal of excavated material; filling; surface treatments.

E10 *In situ concrete*: foundations; beds; slabs; walls; beams and casings; columns and casings; staircases; upstands; grouting; filling.

F10 *Brick/block walling:* walls; isolated piers; isolated casings; chimney stacks; boiler seatings; flue linings; closing cavities; facework ornamental bands; facework quoins; other facework components; bonding to existing; surface treatments.

G20 *Carpentry/timber framing/first fixing:* trusses; trussed rafters/beams; wall or partition panels; portal frames; floor members; wall or partition members; roof members; joist strutting; supports; gutter/fascia/eaves/verge soffit boards; cleats; ornamental ends; wrot surfaces; metal components.

H60 *Clay/concrete roof tiling:* roof coverings; wall coverings; abutments; eaves; verges; ridges; hips; vertical angles; valleys; fittings; holes.

L20 *Timber doors/shutters/hatches:* doors; roller shutters and collapsible gates; sliding/folding partitions; hatches; strong room doors; grilles; door frames and door lining sets; bedding and pointing frames.

M40 *Stone/concrete/quarry/ceramic tiling/mosaic:* walls; ceilings; isolated beams/columns; floors; treads; sills; risers; strings; aprons; linings to channels; skirtings; kerbs; corner pieces; accessories.

M60 *Painting/clear finishing:* general surfaces; glazed windows and screens/sash windows/doors; structural metalwork; radiators; railings, fences and gates; gutters; services; coloured bands for coding service pipes.

P20 *Unframed isolated trims/skirtings/sundry items:* skirtings, picture rails, architraves and the like; cover fillets, etc.; isolated shelves and worktops; window boards; unframed pinboards; duct covers; isolated handrails and grab rails; backboards, etc.

R12 *Drainage below ground:* excavating trenches; items extra over excavating trenches; disposal of water; beds/haunchings/surrounds; vertical casings; pipes; items extra over pipes; pipe accessories; manholes; inspection chambers; soakaways; cesspits; septic tanks; preformed systems; connecting to local authority's sewer; testing and commissioning; preparing drawings.

General Rules of Abstracting

It is most important that the entries in the abstract should be well spaced and it is necessary for the surveying assistant or technician doing the working up to look through the dimension sheets, before he starts abstracting, in order to determine, as closely as possible, how many abstract sheets will be required. In example XXVI some of the items are rather closer together than is desirable, in order to conserve paper and keep down the cost of this student textbook.

The items will be entered in the abstract in the same order as they will appear in the bill, as far as practicable, since the primary function of the abstract is to classify and group the various items preparatory to billing, and to reduce the dimensions to the recognised units of measurement. Descriptions are usually spread over two columns with the appropriate dimension(s) in the first column and any deductions in the second column. The total quantity of each item is reduced to the recognised unit of measurement such as the m, m^2, m^3 or tonne.

It is good practice to precede each description in the abstract with the prefix C, S, L or Nr, denoting that the item is cubic, square, linear or enumerated, and this procedure reduces the risk of errors arising with regard to units or quantities.

As to the order of items in each sub-section of the abstract, the usual practice is to adopt the order of cubic, square, linear and finally enumerated items, with labour items preceding labour and materials, smaller items preceding larger ones and cheaper items preceding the more expensive in each group, but it may be necessary to vary this to follow the sequence in SMM7.

Where it is necessary to abstract a number of similar items but of varying sizes, the best procedure is to group these items under a single heading with each size entered in a separate column, as shown in the following example.

L/Pipes in trs., BS clay to BS65, Pt2 & jtd. w. flex. mech. jts.

100 dia.	150 dia.	225 dia.
45.63 (4) 38.60 (5) 15.40 (5) 52.32 (6)	32.45 (3) 26.28 (3)	35.30 (3)

The number in brackets after each dimension represents the page number of the dimension sheet from which the dimension has been extracted, for ease of reference.

Where similar items of varying size are encountered they can sometimes conveniently be entered on the abstract sheet in the following way.

Nr/ E.O. cop., made bend, size: 15	pipe for nom.		
1 (28) 2 (29) 1 (30)	Nr/Ditto, size: 22.	nom.	
	2 (30) 1 (31) 1 (32)	Nr/Ditto, size: 38.	nom.
		2 (33) 1 (34)	

Deductions are entered in the second column under the main heading of the item under consideration, as illustrated in the following example.

S/Bk. wall, thickness: 102.5, comms. in stretcher bond in g.m. (1:1:6).			
95.64 (13)	Ddt		
152.36 (14)	6.95 (53)		
76.19 (16)	8.91 (56)		
8.32 (18)	26.45 (62)		
2.40 (70)	42.31		
334.91			
42.31			
292.60			
= 293 m²			

In the last example the area is expressed in m², as the correct unit of measurement for this class of work. Futhermore, the dimensions have been crossed through to indicate that the deduction has been made and the adjusted area transferred to the bill.

When measuring some items such as glass in panes ≤ 0.15 m² and iron drain pipes in runs ≤ 3.00 m in length, the number of items involved has to be stated in the billed description. A convenient method of dealing with this type of item in the abstract is indicated in the following example.

S/Glazg.w.stand. plain glass, in panes area ≤ 0.15 m², c.s.g. (O.Q.) 3 th. to BS 952 & glazg. to wd. w. l.o. putty.			
18 = 1.47 (48)	(18 panes with a total area of 1.47 m²)		
54 = 4.02 (49)	(54 panes with a total area of 4.02 m²)		

When enumerated items are to be written in the bill following the associated linear item, such as mitres and fitted ends with larger section hardwood skirtings and the like, the best method of dealing with them in the abstract is as follows.

L/sktg. 38 × 225 wrot. hwd. mo., fxd. w. grds. plugd. to bwk.			
85.20 (38)	Nr/*Ls*		
	16 (38)	Nr/Fitted ends	
		12 (38)	

On completing the entry of all items on the abstract, all entries will be checked, columns of figures cast, deductions made, totals reduced, all the latter work checked and the totals finally transferred to the bill.

BILLING

Ruling of Bill of Quantities

It is desirable that one of the rulings detailed in the British Standard Specification *Stationery for Quantity Surveying* (BS 3327: 1970) should be used to ensure that a uniform method of setting out the information in a bill of quantities is obtained.

This British Standard specifies the rulings for both single and double bill papers (with one or two sets of pricing columns.) The single bill with right-hand billing is the most widely used for building as well as civil engineering work.

The single bill paper as prescribed by BS 3327 is now illustrated. The widths of columns vary slightly on the face and reverse sides of each bill sheet.

1	2	3	4	5	6	7

Nr of column	*Use of column*	*Width (mm) face side*	*reverse side*
1	Item nr	19	14
2	Description	100	100
3	Quantity	24	24
4	Unit	14	14
5	Rate (£)	18	18
6	£	21	21
7	p	14	19

The double billing paper with two sets of pricing columns, is mainly used for bills of variations, one set of columns being used for omissions and the other set for additions. It can also be used in demolition and alteration work where the additional pricing column can be used for the insertion of credits for old materials which are to become the property of the contractor (SMM C10.D1).

Referencing of Items

It is essential that items in a bill of quantities which are to be priced by a contractor shall be suitably referenced. With bills of quantities for building works a common practice is to letter the items alphabetically on each page to avoid the use of the large numbers which arise if all the items are numbered consecutively throughout the bill. Thus the third item on page 20 of the bill of quantities could be referred to as item 20/C (page 20, item C). Another alternative would be to use the SMM references but these are excessively long and complicated.

Entering Quantities in the Bill

When transferring quantities to the bill in metres, they are to be billed to the nearest whole unit. Fractions of a unit which are less than one-half are disregarded and all other fractions (one-half or over) are taken as whole units. Where the unit of billing is the tonne, quantities shall be billed to the nearest two places of decimals.

Where the application of this principle would cause an entire item to be eliminated, the item is to be given as one unit (SMM General Rules 3.3).

Units of Measurement

The words used in describing work of one, two or three dimensions are linear, square and cubic respectively. These words are now little used in practice and the following abbreviations are given in SMM7: metre(s): m; square metre(s): m^2; cubic metre(s): m^3; tonne(s): t; enumerated items: nr (SMM General Rules 12.1).

General Rules of Billing

Example XXVII, included later in this chapter, incorporates the billed items for the foundations of a small building, based on the entries in the abstract forming Example XXVI. As each item is transferred to the bill it is crossed through on the abstract to prevent any possibility of errors occurring during the transfer stage.

The bill columns are headed with their functions in the examples in this chapter, mainly for the guidance of the student, but in practice they are often omitted.

The order of billed items will be the same as in the abstract, as far as practicable, and they will be grouped under suitable work section and sub-section headings as described earlier in the chapter. The work section headings will generally follow the order and terminology adopted in the *Standard Method*, such as Excavating and Filling and *In situ* Concrete. There will usually be a number of preamble clauses inserted at the head of each work section, relating to financial aspects of the work in the section concerned and giving guidance to the contractor in his pricing of items. In addition preamble clauses are frequently used to give detailed material and workmanship requirements with a view to reducing the length of subsequent billed descriptions and eliminating the specification, as illustrated later in the chapter. However, where there is a separate project specification, the preamble clauses can be omitted and there will be extensive cross referencing to the specification in the descriptions of the billed items.

Each item to be priced in the bill is indexed by letters and/or numbers in the first column. It will be noticed that all words in the billed descriptions are written in full wihout any abbreviations and this procedure should always be followed to avoid any possible confusion arising.

Provision is generally made for the total sum on each page of the bill relating to a given section of work to be transferred to a collection at the end of the work section. The totals of each of the collections are transferred to a summary at the end of the bill, the total of which will constitute the contract sum. This procedure is preferable to carrying forward the total from one page to another in each work section, since the subsequent rectification of pricing errors may necessitate alterations to a considerable number of pages.

Billed descriptions should conform to the requirements of the *Standard Method*, follow in a logical sequence and be concise, yet must not, at the same time, omit any matters which will be needed by the contractor if he is to be able realistically to assess the price for each item.

The following example illustrates the normal method of entering items which are 'written short' with the descriptions set back, and the contractor is able to price the enumerated items immediately after the linear item with which they are associated.

F	Skirting 25 × 150 wrought hardwood, moulded, screwed with brass cups and screws and plugged to masonry.	71	m			
G	Mitres	16	nr			
H	Fitted ends	12	nr			
J	25 × 200 ditto.	22	m			
K	Mitres	8	nr			
L	Fitted ends	4	nr			

Note the use of the word 'ditto' when a similar item occurs to avoid unnecessary repetition of descriptions. In practice the words 'wrought hardwood' might be omitted from the description of the skirting and included in a heading to this sub-section of the bill.

On completion the draft bill must be very carefully checked against the abstract and the abstract suitably marked in coloured ink or pencil as each item is dealt with. Particular care should be taken to ensure that all the quantities, units and descriptions are correct that proper provision has been made for section and sub-section headings, transfer of totals to collections and summary and a satisfactory sequence of items obtained.

Further checking arises in connection with the printer's proof which must be carried out extremely carefully. It is also good policy to calculate the approximate areas and volumes of major items of work such as excavation, disposal of excavated material, brick and block walling, roof coverings, painting and also the total number of fittings such as number of windows, doors, sanitary appliances and manholes and to compare them with the actual billed quantities, to ensure that no major errors have occurred.

Preliminaries Bill

The first sectional bill in a bill of quantities is often termed the 'Preliminaries Bill' or 'Preliminaries and General Conditions Bill' and covers many important financial matters which relate to the contract as a whole and are not confined to any particular work section, and the contractor is thereby given the opportunity to price them. Section A of the *Standard Method* details most of the preliminaries and general conditions items which would appear in such a bill.

SMM7 introduced significant changes in the rules for preliminaries. Employer's requirements and limitations (SMM A30–37) are now clearly separated from the contractor's general cost items for management and staff, site accommodation, services and facilities, mechanical plant and temporary works (SMM A40–44).

Thus the preliminaries section of a bill of quantities contains two separate and distinct types of item, and these are now outlined.

(1) Items which are not specific to work sections but which have an identifiable cost which is best considered separately for tendering purposes, such as contractual requirements for insurances, site facilities for the employer's representative and payments to the local authority.
(2) Items for fixed and time related costs which derive from the contractor's expected method of carrying out the work, such as bringing plant to and from the site, providing temporary works and supervision.

The fixed and time related subdivision given for a number of preliminaries items enables tenderers to price the elements separately should they so desire. Tenderers also have the facility at their discretion to extend the list of fixed and time related charges to suit their particular methods of construction.

The first four items in work section A of SMM7 (Preliminaries/General Conditions) relate to the project particulars, list of drawings from which the bills of quantities were prepared, details of the site and existing buildings and services and a description of the work (SMM A10–13). In this way the contractor is able to quickly obtain a general impression of the project and its main implications.

The next item contains a schedule of clause headings of the standard conditions of contract, any special conditions or amendments, appendix insertions, employer's insurance responsibility and performance guarantee bond where applicable (SMM A20).

Items SMM A30–37 prescribe in detail the employer's requirements or limitations ranging from tendering and sub-letting to documents; management of the works; quality standards and control; security, safety and protection; method, sequence and timing; facilities, temporary work and services; to operation and maintenance of the finished building. There is provision for the separation of all these items into fixed and time related charges.

The contractor is given the opportunity to price a wide range of general cost items in SMM A40-44, to which he can add such further items as he wishes and to separate them into fixed and time-related charges. For example, the schedule of mechanical plant items in SMM A43.1.1–8, ranging from cranes to concrete and piling plant, provides a useful check list for pricing purposes. Some items can appear either as employer's or as contractor's requirements, since work such as temporary hoardings may on occasions be fully defined by the tender documents and in other instances be left to the contractor's discretion.

Items SMM A50–55 cover work undertaken or materials supplied by the employer; nominated sub-contracts containing descriptions of the work, provision for main contractor's profit and details of any special attendance required; details of materials to be supplied by nominated suppliers with provision for main contractor's profit; work by statutory authorities; provisional work and dayworks. A single item is provided in the preliminaries bill for general attendance on all nominated sub-contractors and suppliers. Prime cost and provisional sums are described in chapter 2 and dayworks are covered later in this chapter.

Bills of Reduction and Addenda Bills

Bills of Reduction or Reduction Bills are required when the tenders exceed the estimate and the employer requires the tender figure reduced in amount.

To accomplish this, changes often have to be made in the size and/or specification of the works and the quantity surveyor has to prepare a further bill in which some of the more expensive items may be replaced by cheaper items and possibly the dimensions of the building(s) reduced. Thus the wall/floor ratio of the building might be reduced, internal partitions reduced from 100 m to 75 mm in thickness and cheaper facing bricks and internal finishings substituted.

Addenda Bills cover additional work which is added to a contract after the main bills have been prepared. They may cover such items as additional supplementary buildings or increased car parking facilities.

Bills of Reduction give the details of variations and incorporate omissions and additions to the original measured work; details of which can be billed in one of two ways. One method is to to use single bills picking up all the omitted items to start with and then following with the replacement items. Alternatively a double bill with two sets of pricing columns can be used, in which the first set of pricing columns will take the omissions, with the additions inserted in the second set of pricing columns. The examples on pages 242–245 will serve to illustrate both of these methods.

Reduction Bill of Quantities

(1) Substitution of 150 mm bed of hardcore in lieu of 175 mm bed.
(2) Painting brick walls and concrete block partitions with two coats of emulsion paint in lieu of three coats of oil paint.

Abstract prepared from dimensions in Example I. EXAMPLE XXVI

EXCAVATING AND FILLING (D20) 1

PROJECT TITLE:

(Note: the abstract is normally prepared across an A3 sheet)

Excavation		Earthwork Support		Disposal of Excavated Material	
s/Exc. topsoil for preservtn. av. 150 dp.	Ddt.	s/Earthwk. support max. depth ≤ 1.00 m, dist. between opposg. faces ≤ 2.00 m.		c/Disposal of excavtd. mat. off site.	
31·90 (1)	3·65 (1)	32·76 (2)		3·38 (2)	
3·65		= $33 m^2$		2·70 (3)	
28·25				6·08	
= $28 m^2$				= $6 m^3$	
c/Exc. tr. width > 0·30 m, max. depth ≤ 1·00 m.				c/Disposal of excvtd. mat. on site in spoil heaps av. dist. of 30·00 m from excavn.	Ddt.
11·01 (1)				4·79 (1)	0·54 (1)
= $11 m^3$				0·54	
				4·25	
				= $4 m^3$	
		Disposal of Surface Water		**Filling**	
		Disposal of surf. water.		c/Fillg. to excavns. av. thickness > 0·25 m. arisg. from excavns.	Ddt.
		Item (4)		11·01 (1)	3·38 (2)
				6·08	2·70 (3)
				4·93	6·08
				= $5 m^3$	
				c/Fillg. to excavns., av. thickness ≤ 0·25 m from on site spoil heaps, topsoil.	
				0·81 (4)	
				= $1 m^3$	
				Surface Treatment	
				s/Compactg. bott. of excavn.	
				14·69 (2)	
				= $15 m^2$	

Abstract (contd.)

IN SITU CONCRETE (E10) 2

PROJECT TITLE:

In situ Concrete

c/In situ conc. fdns. (1:3:6/40 agg.) poured on or against earth.

3·38 (2)
= 3 m^3

c/In situ conc. (1:6) fillg., holl. wall, thickness ≤ 150.

0·67 (3)
= 1 m^3

Abstract (contd.)

BRICK/BLOCK WALLING (F10) 3

PROJECT TITLE:

Brick walls		
S/ Bk. wall, thickness: 102·5, comms. in stret. bond in c.m. (1:3).		
32·50 (2) 4·64 27·86 = 28m²	Ddt. 4·64 (4)	
S/ Bk. wall, facewk. o.s., thickness 102·5, in 'X' multi. col. fcg. bks. basic price £215/1000 (delvd. to site) in stret. bond in c.m. (1:3) & ptg. w. nt. flush jt. as wk. proceeds.		Damp-proof courses
4·64 (4) = 5m²		S/ Dpc, width ≤ 225, hor., single layer of hessian base bit. felt to BS 743 ref. A, & bedded in c.m. (1:3).
S/ Form cav. in holl. wall, width: 50, inc. 4 nr. wall ties/m² of zinc coated m s. vert. twist type to BS 1243.		3·92 (3) = 4 m²
16·25 (3) = 16m²		

A.K

EXAMPLE XXVII — BILL OF QUANTITIES
(prepared from abstract in Example XXVI)
SMALL BUILDING

Item nr	*Description*	*Qty.*	*Unit*	*Rate*	£	p
	BILL Nr 2 (D20) EXCAVATING AND FILLING					
	Preamble clauses covering nature of ground, ground-water level and other site characteristics.					
	Excavation					
A	Excavate topsoil for preservation average 150 deep.	28	m^2			
B	Excavate trench, width > 0.30 m, maximum depth ≤ 1.00 m.	11	m^3			
	Earthwork Support					
C	Earthwork support, maximum depth ≤ 1.00 m, distance between opposing faces ≤ 2.00 m.	33	m^2			
	Disposal of water					
D	Disposal of surface water.	Item				
	To Collection			£		

Note: Bill Nr 1 would be a Preliminaries Bill.
The numbers in brackets after the bill nr represent the SMM7 work section reference.

Bill of Quantities (continued)

Item nr	*Description*	*Qty.*	*Unit*	*Rate*	£	p
	BILL Nr 2 (contd)					
	Disposal of Excavated Material					
A	Disposal of excavated material off site.	6	m^3			
B	Disposal of excavated material on site in spoil heaps, average distance of 30.00 m from excavation.	4	m^3			
	Filling					
C	Filling to excavations, average thickness > 0.25 m, arising from excavations.	5	m^3			
D	Filling to excavations, average thickness ≤ 0.25 m, from on site spoil heaps, topsoil.	1	m^3			
	Surface Treatment					
E	Compacting bottom of excavation.	15	m^2			
	To Collection			£		
	Collection					
	From page 238					
	From page 239					

Note: Where the quantities in any work section or bill extend over more than one page, the need for a Collection arises, to combine the individual page totals preparatory to transfer to the Summary at the end of the document.

Bill of Quantities (continued)

Item nr	*Description*	*Qty.*	*Unit*	*Rate*	£	p
	BILL Nr 3 (E10)					
	IN SITU CONCRETE					
	Preamble clauses covering materials, mixes, tests and the like.					
	In situ Concrete					
A	In situ concrete foundations (1:3:6/40 aggregate) poured on or against earth.	3	m^3			
B	In situ concrete (1:6) filling, hollow wall, thickness ≤ 150.	1	m^3			
	To Summary			£		

Bill of Quantities (continued)

Item nr	*Description*	*Qty.*	*Unit*	*Rate*	*£*	*p*
	BILL Nr 4 (F10)					
	BRICK/BLOCK WALLING					
	Preamble clauses covering materials and other relevant matters.					
	Brick Walls					
A	Brick wall, thickness: 102.5, in commons in stretcher bond in cement mortar (1:3).	28	m^2			
B	Brick wall, facework one side, thickness: 102.5, in × multicoloured facing bricks, basic price £215/1000 (delivered to site) in stretcher bond in cement mortar (1:3) and pointing with neat flush joint as work proceeds.	5	m^2			
C	Form cavity in hollow wall, width: 50, including 4 nr wall ties per m^2 of zinc coated mild steel vertical twist type to BS 1243.	16	m^2			
	Damp-proof Course					
D	Damp-proof course, width ≤ 225, horizontal, of single layer of hessian base bitumen felt to BS 743 ref. A and bedded in cement mortar (1:3).	4	m^2			
	To Summary			£		

EXAMPLE XXVIII

Using Single Bill *Reduction Bill of Quantities*

Item	*Description*	*Qty.*	*Unit*	*Rate*	£	p
	OMISSIONS					
	Excavating and Filling					
A	Excavate topsoil for preservation average 175 deep.	2550	m^2			
B	Disposal of excavated material on site in spoil heaps, average distance of 30.00 m from excavation.	446	m^3			
C	Filling to make up levels, average thickness $\leq$ 0.25 m, obtained off site, selected gravel rejects.	406	m^3			
	Painting					
D	Painting general surfaces, seal and apply three coats of hard gloss paint on fair faced masonry.	3060	m^2			
	Total of Omissions carried to Collection			£		
	ADDITIONS					
	Excavating and Filling					
A	Excavate for topsoil for preservation average 150 deep.	2550	m^2			
	Carried forward			£		

Using Single Bill (continued) *Reduction Bill of Quantities*

Item	*Description*	*Qty.*	*Unit*	*Rate*	£	p
	ADDITIONS (contd)					
	Brought forward			£		
B	Disposal of excavated material on site in spoil heaps, average distance of 30.00 m from excavation.	382	m^3			
C	Filling to make up levels, average thickness≤ 0.25 m, obtained off site, selected gravel rejects.	348	m^3			
	Painting					
D	Painting general surfaces, seal and twice emulsion paint fair faced masonry.	3060	m^2			
	Total of Additions carried to Collection			£		
	Collection					
	Omissions					
	Additions					
	Net Omissions – carried to Summary			£		

Using Double Bill *Reduction Bill of Quantities*

Item	Description	Qty.	Unit	Omissions Rate	Omissions £	Omissions p	Additions Rate	Additions £	Additions p
	Excavating and Filling Omissions								
A	Excavate topsoil for preservation average 175 deep.	2550	m^2						
B	Disposal of excavated material on site in spoil heaps, average distance of 30.00 m from excavation.	446	m^3						
C	Filling to make up levels, average thickness ≤ 0.25 m, obtained off site, selected gravel rejects.	406	m^3						
	Additions								
D	Excavate topsoil for preservation average 150 deep.	2550	m^2						
E	Disposal of excavated material on site in spoil heaps, average distance of 30.00 m from excavation.	382	m^3						
F	Filling to make up levels, average thickness ≤ 0.25 m, obtained off site, selected gravel rejects.	348	m^3						
	Carried forward			£					

Using Double Bill (continued) *Reduction Bill of Quantities*

Item	*Description*	*Qty.*	*Unit*	*Omissions* Rate	£	p	*Additions* Rate	£	p
	Brought forward			£					
	Painting Omissions								
A	Painting general surfaces, seal and apply three coats of hard gloss paint on fair faced masonry.	3060	m^2						
	Additions								
B	Painting general surfaces, seal and twice emulsion paint fair faced masonry.	3060	m^2						
	Net Omissions – carried to Summary			£					

Note: Item references and page numbers of the original billed items would often be used in practice to reduce the length of descriptions in the reduction bill of quantities.

GENERAL SUMMARY

Where the bill of quantities is annotated in the SMM7 work sections, the General Summary can take one of two different forms. The simplest approach is to provide monetary collections at the end of each group of work sections, such as D: Groundwork and F: Masonry, the totals of which are then transferred to the General Summary. An example of this format of General Summary is illustrated and shows the discontinuity in the groups of work section references which some surveyors could find off-putting after being accustomed to using a numerical sequence of bill numbering.

The second approach is to provide collections at the end of each work section with the totals transferred to the General Summary. This makes for a very long and fragmented General Summary as, for example, Group E (*In situ* concrete) could be subdivided into a number of separate work sections, such as E10 *In situ* concrete; E20 Formwork for *in situ* concrete; E30 Reinforcement for *in situ* concrete; E40 Designed joints in *in situ* concrete; and E41 Worked finishes to *in situ* concrete. This arrangement is unlikely to be very popular in practice.

GENERAL SUMMARY

	page	£	p
A Preliminaries/General Conditions (including Dayworks, PC and Provisional Sums)			
D Groundwork			
E *In situ* concrete/Large precast concrete			
F Masonry			
G Structural/Carcassing metal/timber			
H Cladding/Covering			
J Waterproofing			
K Linings			
L Windows/Doors/Stairs			
M Surface finishes			
N Furniture/Equipment			
P Building fabric sundries			
Q Paving/Planting/Fencing			
R Disposal systems			
S Piped supply systems			
T Mechanical heating			
U Ventilation			
V Electrical power/lighting			
TOTAL CARRIED TO FORM OF TENDER			

Preambles

The purpose and nature of preamble clauses at the head of each work section in the bill were described in chapter 2. It would perhaps be helpful to the student if a few typical clauses were provided and the Excavating and Filling Bill has been selected for this purpose.

A. Trial holes have been excavated in the positions indicated on Location Drawing G.16, on which are also shown the levels and details of soil, which consist of an average of 225 of topsoil overlying well graded sand. Groundwater level was 295.40 AOD on 15 March 19–.

B. The quantities of excavation, including disposal of excavated material have been measured nett and the Contractor is to allow in his billed rates for any increase in bulk and for any additional excavation that may be required for working space, other than that provided in accordance with the Standard Method of Measurement of Building Works.

C. Excavation rates shall include for grubbing up and removing any tree roots and similar obstructions that may be encountered.

D. Excavations shall be inspected and approved by the Architect and the representative of the Local Authority before any concrete or filling is laid, and any variations measured by the Quantity Surveyor.

E. If any excavations are carried out to a greater width or depth than directed or required, then the Contractor is to make up or fill in to the required width or depth with concrete (1:12) at his own expense, where directed by the Architect, and no payment will be made for the additional excavation.

F. Hardcore shall be clean, hard brick or stone broken to pass a 75 mm ring and be well graded and compacted in layers not exceeding 150 mm thick, with a 800 kg vibrating roller running at a slow speed with a minimum of three passes over each layer.

Note: This is a selection of clauses which are not necessarily comprehensive and they will vary from one contract to another.

Daywork

For the valuation of variations which cannot properly be measured and valued at billed rates or rates derived from them, the Contractor will normally be allowed to charge daywork in accordance with Clause 13.5.4.1 of the JCT Conditions of Contract.

The basis of charging is normally the prime cost of such work as calculated in accordance with the *Definition of Prime Cost of Daywork carried out under a Building Contract* together with percentage additions to each section of the prime cost at the rates inserted by the Contractor in the Bill of Quantities. The rates for plant shall be those contained in the *Schedule of Basic Plant Charges*.

The Contractor will therefore add to the provisional sums contained in the Bill the percentage addition that he requires for overheads and profit. The provisional sums and the percentage additions are then monied out and the total carried to the Summary of the Bill of Quantities, in the manner shown in the following Schedule.

Note: This latter work will be undertaken under different conditions to those prevailing during the contract period, since the contractor will have taken his

DAYWORK SCHEDULE

					£	p
Labour						
Provide the Provisional Sum of £1500.00 for Labour.					1500	00
Add for overheads and profit.			%			
Materials						
Provide the Provisional Sum of £1000.00 for Materials					1000	00
Add for overheads and profit.			%			
Plant						
Provide the Provisional Sum of £500.00 for Plant.					500	00
Add for overheads and profit.			%			
The foregoing Provisional Sums and percentage additions apply to daywork ordered by the Architect prior to the commencement of the Defects Liability Period. Should any dayworks be ordered after the commencement of the Defects Liability Period it is proposed that the Definition shall also apply to such work and the Contractor is invited to insert the percentage additions he would require to the Prime Cost for overheads and profit.						
Labour			%			
Materials			%			
Plant			%			
DAYWORK CARRIED TO SUMMARY			£			

workforce, materials and plant off the site following practical completion of the work. Hence it is likely that he will wish to insert higher percentage rates than those in the main part of the Schedule.

Specialist Bills

On occasions, specialist work such as an electrical installation, arising from a prime cost sum in the main bill, may be measured in detail and the main contractor may be given the opportunity to price this work in addition to specialist sub-contractors. The arrangement of the specialist bill is similar to a normal bill with comprehensive preliminary and preamble clauses covering such matters as the relationship of the specialist to the main contractor, the provision of general services and facilities by the main contractor, details of special attendance, programme and method of payment.

RECENT DEVELOPMENTS IN BILL PREPARATION

New measurement and processing techniques hae been introduced in recent years and they are now being used to an increasing extent, since they are resulting in a speeding up of working up operations and a reduction in the overall cost of preparing bills of quantities.

Over the years many quantity surveyors have experimented with a number of systems designed to eliminate part of the working up process. One such system eliminated the abstract by billing direct as described earlier in this chapter.

Another method was to abstract from standard dimensions paper on to specially ruled sheets or cards, designed to receive only one item per sheet or card. Where repeat items occurred the quantity was recorded with the original item, enabling the total quantity to be obtained eventually. As the abstract was prepared the sheets or cards were sorted into bill order. On completion of the abstract, any necessary editing was done and the bill typed direct from the abstract, resulting in the elimination of both billing and the checking of the bill.

Each of these systems suffered some limitations in use and could not therefore be applied universally.

Cut and Shuffle

General Arrangements

The system of 'cut and shuffle' was developed in the early 1960s and by the late 1970s was probably one of the most widely used methods of entering dimensions and descriptions. It has been aptly described as 'rationalised traditional' procedure. Unlike abstracting and billing there is no universally accepted format and many different paper rulings and methods of implementation are used in different offices. Some offices even use more than one system to suit different types of work.

However, the following characteristics apply to most systems.

(1) Dimensions paper is sub-divided into four or five separate sections which can subsequently be split into individual slips.
(2) Only one description with its associated dimensions is written on each section.
(3) Dimension sheets are subsequently split into separate slips and sorted into bill work sections or elements and eventually into bill order.
(4) Following the intermediate processes of calculation and editing, the slips form the draft for typing the final bill of quantities.

The cut and shuffle system is designed to eliminate the preparation and checking of the abstract and draft bill. Hence there is only one major written operation, namely taking off, compared with the three involved with abstracting and billing.

Detailed Procedure

One method of carrying out the system is described in detail on page 254, with the procedure illustrated diagrammatically in Fig. XX.

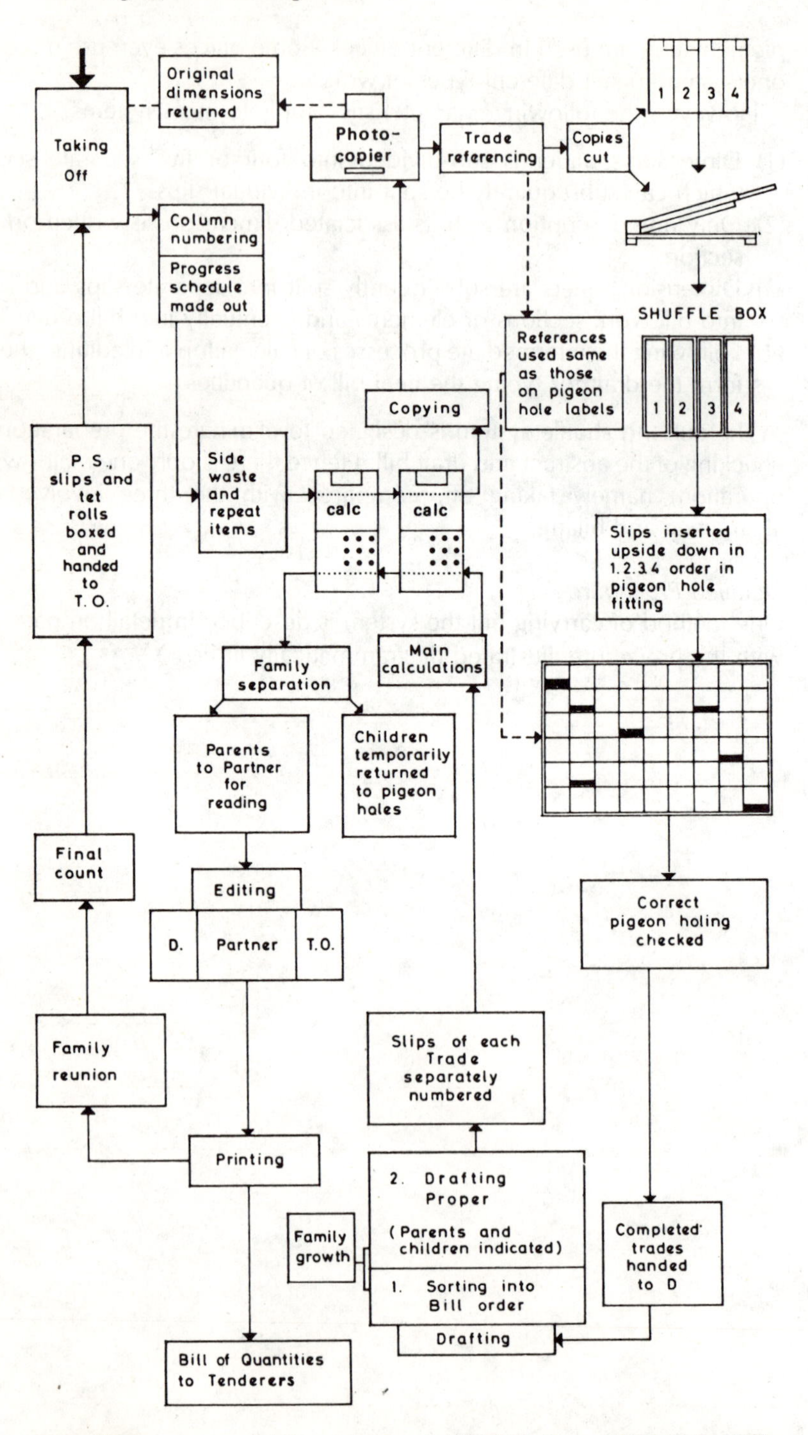

FIG XX CUT AND SHUFFLE PROCEDURE
(Source: B. D. Henderson, Cut and Shuffle System paper, RICS Annual Conference, 1961)

EXAMPLE XXIX

Sheet 233 — CHARLSWTH HOSPITAL | 485 | J40·1 | Flrs.

Damp proofing, horizontal, polythene 1000 gauge laid on concrete bed.

Dimensions	Description
4·50 3·80	(Ward A
20·00 6·00	(do.
5·50 2·00	(Pass.
23·80 6·00	(Ward B
5·50 3·80	(do.
7·60 2·00	(Pass.

233

Sheet 234 — CHARLSWTH HOSPITAL | 485 | J40·2 | Flrs.

Damp proofing a.b. (col. 233)

Dimensions	Description
20·00 6·00	(Ward C
5·50 4·20	(do.
7·60 2·00	(Pass.

234

Sheet 235 — CHARLSWTH HOSPITAL | 485 | M10·1 | Flrs.

Screed to floor, level, thickness : 25, one coat, floated, cement and sand (1:3) on concrete bed (measured separately).

Sq. — J40.1 & 2

235

Sheet 236 — CHARLSWTH HOSPITAL | 485 | M42·1 | Flrs.

Wood block flooring, level, 300x75x25 European Oak dipped and jointed in hot bitumen on floated screed (measured separately), dowelled, laid herringbone, with two block plain margins all round, and sealing and polishing with two coats of wax polish.

Sq. — J40.1 & 2

236

Note: The total area will probably be calculated on a calculating machine and entered on sheets 234, 235 and 236; when the dimensions on sheets 233 and 234 will be crossed through. Each section of each dimension sheet has the project number (485), SMM work section reference and possibly a taking off number, such as J40.1 and J40.2, and sheet number (233–236) entered on it. The latter numbering is continuous throughout the project. The project title (Charlesworth Hospital) will probably be stamped on the sheets with a rubber stamp. No squaring column has been included in this example, as individual calculations can be obtained from the calculator print out.

In practice, many of the currently used cut and shuffle sheets seem to include a squaring column. This makes the checking of the squaring more straightforward and also has benefits when referring back to the original dimensions in the post-contract period.

TYPICAL CUT AND SHUFFLE DIMENSION SHEET

(1) Taking off is carried out on A4 size sheets of dimensions paper, ruled vertically into four sections and thus accommodating four items per sheet. Dimensions are entered on one side only of each sheet and each column is generally stamped with the project reference number and numbered consecutively. Ditto items must include a reference to the column number of the main item, where full particulars can be found. A typical cut and shuffle dimension sheet is illustrated in example XXIX.
(2) As sections of the taking off are completed, the side casts are checked and repeat dimensions calculated.
(3) When a taking off section is complete, each column is marked with the work section reference and column number. A copy of each dimension sheet is obtained generally either by using NCR (no carbon required) paper or by photocopying. However, some systems operate without the need to produce a copy.
(4) The taker off retains the copy and the original sheet is cut into four slips each containing one item. Some surveyors use sheets that are already perforated.
(5) The slips are shuffled or sorted into work sections such as Excavating and Filling and *In situ* Concrete. Similar items are collected together and the whole of the slips placed as near as possible in bill order.
(6) When all the slips for an individual work section have been sorted they are edited to form the draft bill, with further slips being inserted as necessary to provide headings, collections and other relevant items. The correct unit of billing is entered on the 'parent' or primary item slip and the 'children' or repeat item slips are marked 'a.b.' (as before). As each section is edited it is passed to a calculator operator for squaring.
(7) The calculator operator squares, casts, reduces and inserts the reduced quantity on the parent item slip. This operation is then checked.
(8) Parent and children slips are separated. The parent slips form the draft bill and are ready for typing.
(9) Any further checks on the draft bill are then carried out and final copies made and duplicated.
(10) The children slips are then replaced to provide an abstract in bill order for reference purposes during the post-contract period.

The principal merits and demerits of the cut and shuffle system are now listed.

Merits

(1) It is claimed that the time taken from the start of taking off to the production of the finished bill is shorter than by traditional methods; working up time is definitely reduced.
(2) Dimensions and descriptions are only written once and not three times

as with traditional methods, and it also eliminates the preparation and checking of the abstract and billing.

(3) The working up section completes almost the whole process after the taking off stage. As soon as the taking off for each section of work is completed, it can be shuffled, edited, typed, read back and duplicated, so that only a short time after all taking off is completed, the bill is ready for distribution. Previously it was often left to the taker off to bill the abstract after completing the taking off.

Demerits

(1) It is often found that although the time spent in working up is much reduced, the taking off is prolonged. The main reasons for this are the comprehensive system of referencing required on each slip and the need to write full descriptions without abbreviations for all parent items.
(2) After the bill has been produced, the cut and shuffle system is not so well suited to final account work, since without a comprehensive index the location of work can be time consuming.

Standard Descriptions

Since the mid-1960s many quantity surveyors have been using standard descriptions in bills of quantities by arranging the component parts of item descriptions into a graded structure. This followed a recommendation of a working party of the Royal Institution of Chartered Surveyors to develop a logical structure of bill item descriptions to assist the building industry in obtaining a common approved standard.

The basic approach to the graded structure of a billed item description is to adopt a number of levels, each of which contains alternative words or phrases. The following example will serve to illustrate a practical application of this concept, applying SMM7 requirements as far as practicable, although the sequence of items tends to vary with different work sections. Nevertheless, SMM7 if applied systematically as described in chapter 2 and implemented in the worked examples, does produce a standardised approach to the preparation of billed item descriptions, and this can be further assisted by the use of the *SMM7 Library of Standard Descriptions*. The actual wording will vary extensively according to whether or not there is a detailed project specification, entailing the use of numerous cross-references in the bill descriptions.

Level	Content	Example
1	Main section heading (normally a work section in the SMM)	In situ concrete
2	Subsidiary division within the main heading, identifying the work	Beds
3	Details of materials and mix	1:3:6/40 aggregate or 11.50 N/mm^2
4	Size ranges, where applicable	Thickness ≤ 150 mm
5	Other variables, where appropriate	Poured on or against unblinded hardcore

The standard description format would then appear in the bill of quantities in the following way.

Item nr	*Description*	*Qty.*	*Unit*	*Rate*	£	p
	IN SITU CONCRETE 11.50 N/mm^2					
	Beds, poured on or against unblinded hardcore thickness:					
1	≤ 150		m^3			
2	150 – 450		m^3			
3	> 450		m^3			

The use of standard descriptions aids communication through the consistent approach, with resultant benefits to those concerned with the pricing and use of priced bills. The drafting and interpretation of descriptions are simplified and the editing of the draft bill is replaced by routine checks.

Fletcher and Moore developed a standard phraseology to meet the fundamental principles required for a standardised approach to bill descriptions based on SMM 6. It embraced layers or levels of description whereby, by combining words or phrases from each relevant level, ordered descriptions could be compiled for all items of building work. The basic structure of the phraseology was a hierarchical division into work sections, activities, products, constructional features and dimensional factors. The principal aim was to encourage standardisation of terminology, coupled with a systematic presentation of the standard terms, and the system has been widely used in professional offices. In 1986, Fletcher, Moore, Monk and Dunstone produced *Shorter Bills of Quantities: The Concise Standard Phraseology and Library of Descriptions*, as described in chapter 1.

Use of Computers

Computers are being used by quantity surveyors to an ever increasing extent for bill production and other quantity surveying functions, and their main characteristics and uses are now considered.

Computers and Information Technology

The Royal Institution of Chartered Surveyors was investigating the use of computers as an aid to the preparation of bills of quantities as long ago as 1961. In 1971, the Royal Institution declared that "the quantity surveyor, like others in the industry, must be familiar with and learn to use the computer as it opens up new techniques for the more effective practice of his skills," and in 1983 stated that "computer and information technology can aid professional competence." The 1983 report on *The Future Role of The Chartered Quantity Surveyor* described how rapidly advancing technology in the mini/microcomputer field, in word processing and in database information retrieval systems will have tremendous impact on quantity surveying techniques in the next decade, with much improved services to the client.

These developments will assist in producing more accurate assessments of alternative bids in terms of time and cost valued against higher costs and different client's requirements, such as early completion. In addition, the ability to relate funding, cash flow and ordering of resources will result in more sophisticated methods in the financial management of projects becoming available.

Microcomputers and Bill Production

Many makes of microcomputer came on to the market in the 1980s at continually reducing prices. They generally consist of a combined keyboard and monitor (visual display unit or screen), one or two floppy disk (disc) drives and a printer of selected speed and quality. The floppy disks are used extensively as they are relatively cheap and possess quite fast data transfer speeds. The hard disk systems provide increased storage capacity and higher operating speeds. The main disadvantage of microcomputers stems from the incompatibility of much of the equipment, although this problem was being reduced in the mid 1980s. The risk of loss or damage is easily overcome by making duplicate copies of all disks.

An increasing use is being made of microcomputers for the preparation of bills of quantities, resulting from the reduced cost of hardware (equipment) and the improved range and efficiency of software (programs). This has culminated in the provision of a more efficient, improved and faster service. The newer models enable more data to be held on the computer with greater ease of access. An increased number of software packages (programs) are available and these permit the production of bills in different formats.

Computer-aided bill production systems provide the facility to check accuracy, but care is needed in the coding of dimensions and entry of data. Modern computerised billing systems can, however, print out errors in the form of tables. The coding can be double checked, although a random check may be considered adequate. The need to engage outside agencies for computerised bill production has been largely eliminated.

The use of computer-aided bill of quantities production packages eliminates the reducing, abstracting and billing operations by converting coded dimension sheets into bills of quantities. Codes may be based on the SMM7 reference codes to form a standard library of descriptions. Items not covered by the standard library are termed rogue items, but these are unlikely to be very extensive with the increased standardisation and reduction in the number of billed items stemming from the introduction of SMM7. The rogue items are suitably coded and entered into either the standard library or the particular project library. The computer normally prints a master copy of the bill of quantities which can be photostated on to ruled paper to give a high standard of presentation.

Data input systems vary and some offices prefer every taker off to have a work station and to input his own dimensions and descriptions and/or codes as he proceeds. An alternative is for the data to be collected centrally and to be input by a machine operator, thus permitting the taker off to avoid coding if he wishes. It is mainly a question of finance, as to whether it is cheaper for the taker off to spend a little longer and look up his own codes or whether it is better for a lower paid member of staff to do the work. Local circumstances will usually provide the answer, and the size of office and type of workload will have an influence.

There are various types of data input system available. Thus the input method can incorporate traditionally prepared dimensions, suitably coded, or the organisation can use direct keyboard entry with automatic squaring or a fully integrated digitiser.

Range of Microcomputer Programs

Microcomputers can be used to advantage in many activities associated with building projects and the following selection gives an indication of their wide range and scope, stemming from their extensive storage capacity, and ease of retrieval of data and monitoring of progress.

(1) Bills of quantities production
(2) Automatic measurement of some building works
(3) Materials scheduling, possibly linked with computer-aided design (CAD)
(4) Earthwork calculations (cut and fill)
(5) Specification production
(6) Feasibility studies and comparative design cost statements
(7) Cost control
(8) Cost reporting
(9) Estimating and tendering
(10) Tender analysis
(11) Budgetary controls
(12) Valuations
(13) Formula price adjustment
(14) Variations and final accounts
(15) Fee management
(16) Quotations and enquiries
(17) Cash flow forecasting
(18) Capital programming
(19) Project planning and control
(20) Progress statements
(21) Resource analysis
(22) Sub-contractors' payments
(23) Maintenance scheduling
(24) Plant and equipment scheduling.

OTHER BILL FORMATS

This final chapter concludes with a conspectus of the various alternative bill formats that have been introduced in the last two decades, mainly aimed at securing greater value to the contractor by simplifying the pricing of the work at the tender stage and by providing greater benefits at the construction stage.

Elemental Bills

As described in chapter 1, an elemental bill of quantities is divided into appropriate building elements instead of the normal work sections. Hence

Excavating and Filling, *In situ* Concrete and Brick/Block Walling are replaced by such bill headings as Substructure, External Walls, Internal Walls and Floors. It is advantageous to use the RICS Building Cost Information Service (BCIS) elements and sub-elements to secure standardisation and assist with cost planning and cost analysis work. Within each element work may be billed in order of work sections or grouped in building sequence. The principal objective is to secure more precise tendering by making the location of the work more readily identifiable and to provide a closer link with the cost plan.

In practice this bill format has not been very well received, since it has involved considerable repetition of billed items, and where work is let to a sub-contractor it is necessary to prepare an abstract of items prior to obtaining quotations. In like manner an estimator has to examine the elements in some detail to collect together all like items before he can assess the total quantities of each material.

Sectionalised Trade Bills

The sectionalised trade bill was produced to overcome some of the objections to the elemental bill, in that it can be presented either as a work section bill or as an elemental bill. For tendering purposes the work section order is used sub-divided into elements. The document is produced in loose-leaf form with each element on a separate sheet.

The estimator is still faced with a certain amount of repetition at the tender stage. The successful contractor is provided with several copies of the bill, which permits the use of work section order for purchasing materials and elemental order for site management.

Operational Bills

Operational bills were developed by the Building Research Establishment and they sub-divide the work into site operations as distinct from trades or elements. Labour and sometimes plant requirements are described in terms of the operations required, together with a schedule of the materials for each operation. Operations are defined as the work done by a man, or by a gang of men, at some definite stage in the building process. The sequence of operations is often illustrated in the bill by means of a precedence diagram, which shows their interrelationship.

It was considered by the main protagonists of the method that the contractor could more readily appreciate the implications of the design at the time estimates were prepared, and that the more detailed information would be of considerable value to the successful contractor on the site, without curtailing the contractor's freedom to select the best constructional techniques.

It involved a fundamental divergence from a traditional bill, with the separation of labour and material items and required significant changes in the rules of measurement prescribed in the *Standard Method*. The whole of one work section will not appear together and so it is customary to print each work section on differently coloured paper to facilitate the location of items.

Operational bills are both bulky and costly to produce. On the other hand, typing and printing of the bill can be started at an earlier stage and there is a reduction in the amount of editing necessary. It also simplifies the work of making interim valuations. They generally increase the work of the contractor's estimating department, although the estimator obtains a clearer picture of the work involved and this should enable him to produce more realistic estimates, and also to ease considerably the task of obtaining quotations for materials. From the site organisational viewpoint, the contractor is helped since his programme is linked automatically with the bills and his manpower requirements are readily assessed from the operations.

In practice the response to this particular bill format was very disappointing. The need for a complete set of drawings prior to the preparation of the bills, coupled with a delayed start on the site, may have been felt to outweigh the possible advantages of a faster and cheaper completion. There was also the natural reluctance to make such a radical change from traditional methods, and it would not conform to normal estimating practice.

Activity Bills

The activity bill is a development of the operational form but without the separation of labour and materials. It is sub-divided into sections based on activities or operations derived from a network analysis. The work is measured in accordance with the *Standard Method*, although on site and off site activities are usually separated and special equipment and components and the work of nominated specialists may be grouped in separate bills.

Some activities need to be completed before others can be started. For example, foundation trenches need to be excavated before the concrete foundations can be laid, which must in turn precede the building of brickwork. A graphical representation in the form of a network analysis shows the interrelationship and sequence of activities. Network analyses for the smaller and less complex projects can be produced by the design team, but those for larger and more complicated contracts may need liaison with the contractor, which could militate against competitive tendering.

Annotated Bills

It is possible for the bill of quantities to give the contractor full details of the quantity, type and quality of materials and labour, and for an accurate and complete set of drawings to show him precisely where and how the work is to be executed. Nevertheless, there are always some billed items whose location in the works is not readily identifiable, and it is most useful to have a note against them in the bill giving their location. This approach has resulted in the production of annotated bills.

Annotations may be prepared in a separate document from the bill of quantities or they may be bound in at the back of the bill, although in either case they must be carefully cross-referenced. Another alternative is to interleave annotations into the bill so that the notes appear opposite the relevant bill items. This provides the clearest and most helpful method.

General Conclusions

These more specialised types of bill format each have their own particular attributes and, in certain situations, a valuable role to perform in tendering procedures and contract administration. Nevertheless, the traditional work section format remains the one most widely used and is likely to remain so for the foreseeable future.

On the other hand, tendering arrangements have been subject to considerable change away from *open tendering* — previously mainly practised by public authorities who advertised publicly and invited tenders from all respondents — towards *selective tendering* — whereby a limited number of contractors of known ability and financial standing are invited to submit tenders — and *negotiated tendering* — where only one contractor is involved and the benefit of contractor participation at the design stage can be achieved. Within the negotiated tender approach there are a number of alternatives including two stage tendering and continuity contracts, which embrace serial, continuation and term contracts; all of which are described in *Quantity Surveying Practice* by the same author.

The National Joint Consultative Commmittee for Building (NJCC) has advised that contractual difficulties may be encountered where contractors tender for certain categories of works based on drawings and specifications alone, although it is accepted that bills of quantities may not always be warranted for small or relatively simple building works. It is further recommended that where tenders are invited without bills of quantities it may well be desirable instead to require the tender price to be broken down into the major constituent parts for ease of cross checking and comparison with other tenders and to require the pricing of a schedule of rates to assist in the valuation of any variations.

Appendix I List of Abbreviations

a.b. *as before*
a.b.d. *as before described*
abut. *abutment*
adj. *adjoining or adjustment*
a.f. *after fixing or angle fillet*
agg. *aggregate*
agst. *against*
allce. *allowance*
alt. *alternate*
alum. *aluminium*
appvd. *approved*
a/r *all round*
archve. *architrave*
ard. *around*
asp. *asphalt*
assctd. *associated*
art. *artificial*
av. *average*

bast. *basement*
bd. *board*
bdd. *boarded*
bdg. *boarding*
bdy. *boundary*
b.e. *bossed end or both ends*
bearg. *bearing*
benchg. *benching*
b.f. *before fixing*
b.i. *build in or built in*
b.i.g. *back inlet gully*
bit. *bitumen*
b. & j. *bed and joint*
bk. *brick*
bkg. *breaking*
bkt. *bracket*
bldg. *building*
blk. *block*
b.m.s. *both measured separately*
b.n. *bullnose*
bott. *bottom*
b. & p. *bed and point*
br. *branch*
breakg. *breaking*
brr. *bearer*
b.s. *both sides*
BS *British Standard*
b.s.m. *both sides measured*
bwk. *brickwork*

cal. plumb. *calcium plumbate*
calkg. *caulking*
cant. *cantilever*
cap. *capacity or capillary*
cappg. *capping*
casg. *casing*
cast. *casement*
cat. *catalogue*
cav. *cavity*
C.C. *curved cutting*
C.C.N. *close copper nailing*
ccs. *centres*
c. & f. *cut and fit*
chan. *channel*
chfd. *chamfered*
chnlk. *chain link*
chr. *chromium*
chy. *chimney*
c.i. *cast iron*
circ. *circular*
cistn. *cistern*

c.j. *close jointed*
clg. *ceiling*
cln. *clean*
clse bdd. *close boarded*
c.m. *cement mortar*
col. *colour*
comb. *combined*
comm. *commencing*
comms. *commons*
commsng. *commissioning*
comp. *compression*
compactn. *compaction*
compo. *composition*
conc. *concrete*
concld. *concealed*
conn. *connection*
constn. *construction*
cont. *continuous*
contd. *continued*
cop. *copper*
copg. *coping*
cos. *course*
covd. *covered*
coverg. *covering*
c.p. *chromium plated*
cplg. *coupling*
c. & p.. *cut and pin*
crnr. *corner*
c.s.a. *cross sectional area*
csd. *coursed*
c.s.g. *clear sheet glass*
c'sk. *countersunk*
ct. *coat or cement*
cu *cubic*
cultvd. *cultivated*
cupd. *cupboard*
curv. *curved*
cuttg. *cutting*
c.w. *cold water*

dble. *double*
ddt. *deduct*
delvd. *delivered*
dep. *deposit*
dia. or diam. *diameter*
diag. *diagonally*
diagrm. *diagram*
diff. *difference*
dimnsd. *dimensioned*
disch. *discharge*
dist. *distance or distemper*
ditto. or do. *that which has been said before*
div. *divided*
dp. *deep*
d.p.c. *damp-proof course*
dr. *door or drain*
dwg. *drawing*

ea. *each*
earthwk. *earthwork*
edgg. *edging*
e.g. *eaves gutter*
emulsn. *emulsion*
enam. *enamel*
eng. *engineering*
Eng. *English*
e.o. *extra over*
ex. *exceeding or extra*
exc. *excavate*
excavn. *excavation*
exp. *exposed*
ext. *external*

facewk. *facework*
fast. *fastener*
fcg. *facing*
f/cly. *fireclay*
fdn. *foundation*
f.e. *fair end*
f.f. *fair face*
fillg. *filling*
fin. *finish or finished*
firrg. *firring*
fittg. *fitting*

fl. *flush*
Flem. *Flemish*
flex. *flexible*
flg. *flooring*
flgd. *flanged*
flr. *floor*
fltd. *floated*
F.O. *fix only*
follg. *following*
fr. *frame*
frmg. *framing*
ft. *feet*
furn. *furniture*
fwd. *forward*
fwk. *formwork*
fxd. *fixed*
fxg. *fixing*

galvd. *galvanised*
g.b.d.p. *gas barrel distance pieces*
gen. *general*
g.i. *galvanised iron*
g.1. *ground level*
glzd. *glazed*
glzg. *glazing*
g.m. *gauged mortar*
gradg. *grading*
grassld. *grassland*
grd.(s) *ground(s)*
grtd. *grouted*
grtg. *grouting*
grve. *groove*
g.s. *general surfaces*
gtg. *grating*
gth. *girth*
g.w.i. *galvanised wrought iron*
gyp. *gypsum*

H.A. *highway authority*
h.b. *half-brick*
h.b.s. *herringbone strutting*
hd. *hard or head*
hd/rl. *handrail*
h.c. or hdcore. *hardcore*
hdd. *headed*
hdg. *heading*
herrgbone. *herringbone*
hg. *hung*
hi. *high*
h.j. *heading joint*
hkd. *hooked*
holl. *hollow*
hor. *horizontal*
h.p. *high pressure*
hsd. *housed*
h.r. *half round*
ht. *height*
hth. *hearth*
hwd. *hardwood*

inc. *including*
indvdl. *individual*
insulatn. *insulation*
inter. *intermediate*
int. *internal*
invt. *invert*
irreg *irregular*
isoltd. *isolated*

jb. *jamb*
jst. *joist*
jt. *joint*
jtd. *jointed*
jtg. *jointing*
junctn. *junction*

kg *kilogram(s)*
km *kilometre(s)*
k.p.s. *knot, prime and stop*

la. *large*
L.A. *local authority*
lapd. *lapped*
layg. *laying*
ld. *lead*
len. *length*
lev. *level*
lg. *long*

lin. *linear*
ling. *lining*
l.m. *lime mortar*
l.o. linseed oil
l.p. *large pipe*

m *metre(s)*
m^2 *square metre(s)*
m^3 *cubic metre(s)*
mach. *machine*
mat. *material*
max. *maximum*
mech. *mechanical*
med. *medium*
memb. *membrane*
membr. *member*
mesd. *measured*
met. *metal*
m/gd. *make good*
m.h. *manhole*
mi. *mitre*
min. *minimum*
mldd. *moulded*
mm *millimetre(s)*
mo. *moulded or mortar*
mors. *mortice*
m.s. *mild steel*
m/s *measured separately*
mull. *mullion*
multi-col. *multi- coloured*

n.e. *not exceeding*
nec. *necessary*
nr *number*
nld. *nailed*
nom. *nominal*
nsg. *nosing*
nt. *neat*

o/a. *overall*
o.e. *one end or other end*
o/fl. *overflow*
o'hg. *overhang*
o/let. *outlet*
o'll. *overall* (alternative to o/a.)
opg. *opening*
optd. *operated*
O.Q. *ordinary glazing quality*
ord. *ordinary*
o.s. *one side*
oslg. *oversailing*
(3) *three oils*

pan. *panel*
patt. *pattern/patterned*
pavg. *paving*
p.c. *prime cost*
p.d.p. *plastic distance pieces*
ped. *pedestal or pedestrian*
perf. *perforation*
perm. *permanent*
picrl. *picture rail*
pla. *plaster*
pl. blk. *plinth block*
pltd. *plated*
plugd. *plugged*
p.m. *purpose made*
p.o. *planted on*
pol. *polish/polished*
polyth. *polythene*
polyst. *polystyrene*
posn. *position*
ppt. *parapet*
pr. *pair*
prep. *prepare*
preservn. *preservation*
proj. *projecting/projection*
provsn. *provision*
provsnl. *provisional*
p.s. *pressed steel*
P.St. *Portland stone*
p. & s. *plugged and screwed*
pt. *point, paint or part*
ptg. *pointing*
ptn. *partition*

q.t. *quarry tile*
qual. *quality*

rad. *radius*
rakg. *raking*
R.C. *raking cutting*
r. conc. *reinforced concrete*
rd. *round or road*
rdd. *rounded*
rdwy. *roadway*
reb. *rebated*
rec. *receive*
rect. *rectangular*
red. *reduced*
ref. *reference*
reg. *regular*
rf. *roof*
reinfd. *reinforced*
reinft. *reinforcement*
rendg. *rendering*
retn. *return*
rg. *ring*
r. & g. *rubbed and gauged*
rl. *rail*
r.l. *red lead*
r.l.j. *red lead joint*
ro. *rough*
r.o.j. *rake out joints*
r.s. *rolled steel*
r. & s. *render and set*
r.s.j. *rolled steel joist*
rt. *root*
r.w. *rainwater*
r.w.p. *rainwater pipe*

san. *sanitary*
s.c. *stop cock*
scrd. *screwed or screed*
scrdn. *screw down*
sd. *sand*
s.e. *stopped end*
sec. or sectn. *section*
sel. *selected*
serv. *service*
S.F. *stepped flashing*
s.j. *soldered joint*
sk. *sunk/sink*
sktg. *skirting*
shelvg. *shelving*
s.l. *short length*
slopg. *sloping*
sm. *small*
smth. *smooth*
sn. *sawn*
s.n. *swanneck*
soc. *socket*
soff. *soffit or soffite*
s.p. *small pipe*
spec. *special*
spld. *splayed*
sprd. *spread*
sq. *square*
s. & s. *spigot and socket*
st. *stone or straight*
stackg. *stacking*
stand. *standard*
stl. *steel*
stret. *stretcher*
stripd. *stripped*
strng. *straining*
strt. *straight*
struct. *structure*
sty. *storey*
suppt. *support*
surf. *surface*
surrd. *surround*
susp. *suspended*
S. & V. P. *soil and vent pipe*
s.v. *stop valve or sluice valve*
swd. *softwood*

tankg. *tanking*
tarmac. *tarmacadam*
tbr. *timber*
temp. *temporary*

t.c. *terra cotta*
ten. tenon
t. & g. *tongued and grooved*
tgd. *tongued*
th. *thick*
thermp. *thermoplastic*
thro. *throated or through*
thsd. *thousand*
tiltg. *tilting*
tk. *tank*
tog. *together*
tr. *trench*
triang. *triangular*
t. & r. *tread and riser*
trav. *traversed*
trd. *tread*
trimg. *trimming*
trowld. *trowelled*

u/c *undercoat*
u/grd. *underground*
uncsd. *uncoursed*
underclk. *undercloak*
upstd. *upstand*
u/s. *underside*

vert. *vertical*
vit. *vitreous*

wd. *wood*
wdw. *window*
wethd. *weathered*
w.g. *white glazed*
w. *with*
w.i. *wrought iron*
wk. *work*
wkg. sp. *working space*
w.p. *wax polish or waterproof*
W.P. *waste pipe*
wrot. *wrought*
w/s *working space (alternative to wkg. sp.)*
wt. *weight*

x-reb. *cross rebated*
xtg. *existing*
x-tgd. *cross tongued*

Y. st. *York stone*

Note: the abbreviations *SMM* and *SMM7* have been used extensively throughout this book and refer to the *Standard Method of Measurement of Building Works: Seventh Edition.*

Appendix II Mensuration Formulae

Figure	*Area*
Square	$(\text{side})^2$
Rectangle	length × breadth
Triangle	$\frac{1}{2} \times$ base × height or $\sqrt{[s(s-a)(s-b)(s-c)]}$ where $s = \frac{1}{2} \times$ sum of the three sides and a, b and c are the lengths of the three sides.
Hexagon	$2.6 \times (\text{side})^2$
Octagon	$4.83 \times (\text{side})^2$
Trapezoid	height × $\frac{1}{2}$ (base + top)
Circle	$(22/7) \times \text{radius}^2$ or $(22/7) \times \frac{1}{4}\,\text{diameter}^2$ (πr^2) $(\pi D^2/4)$ circumference = $2 \times (22/7) \times$ radius or $(2\pi r)$ $(22/7) \times$ diameter (πD)
Sector of Circle	$\frac{1}{2}$ length of arc × radius
Segment of Circle	area of sector – area of triangle

Figure	*Volume*	*Surface Area*
Prism	area of base × height	circumference of base × height
Cube	$(\text{side})^3$	$6 \times (\text{side})^2$
Cylinder	$(22/7) \times \text{radius}^2 \times \text{length}$ $(\pi r^2 h)$	$2 \times (22/7) \times \text{radius} \times (\text{length} + \text{radius})$ $[2\pi r(h + r)]$
Sphere	$(4/3) \times (22/7) \times \text{radius}^3$ $(4/3\pi r^3)$	$4 \times (22/7) \times \text{radius}^2$ $(4\pi r^2)$
Segment of Sphere	$(22/7) \times (\text{height}/6) \times (3\ \text{radius}^2 + \text{height}^2)$ $[(\pi h/6) \times (3r^2 + h^2)]$	curved surface $= 2 \times (22/7) \times \text{radius} \times \text{height}\ (h)$ $(2\pi rh)$
Pyramid	$\frac{1}{3}$ area of base × height	$\frac{1}{2}$ circumference of base × slant height
Cone	$\frac{1}{3} \times (22/7) \times \text{radius}^2 \times \text{height}$ $(\frac{1}{3}\pi r^2 h)$	$(22/7) \times \text{radius} \times$ slant height (l) (πrl)
Frustum of Pyramid	$\frac{1}{3}$ height $[A + B + \sqrt{(AB)}]$ where A is area of large end and B is area of small end.	$\frac{1}{2}$ mean circumference × slant height
Frustum of Cone	$(22/7) \times \frac{1}{3}$ height $(R^2 + r^2 + Rr)$ where R is radius of large end and r is radius of small end. $[\frac{1}{3}\pi h(R^2 + r^2 + Rr)]$	$(22/7) \times$ slant height $(R + r)$ $[\pi l(R + r)]$ where l is slant height

For Simpson's rule and prismoidal formula see chapter 3.

Appendix III Metric Conversion Table

Length	1 in. = 25.44 mm [approximately 25 mm, then (mm/100) × 4 = in.]
	1 ft = 304.8 mm (approximately 300 mm)
	1 yd = 0.914 m (approximately 910 mm)
	1 mile = 1.609 km (approximately 1 $\frac{3}{5}$ km)
	1 m = 3.281 ft = 1.094 yd (approximately 1.1 yd) (10 m = 11 yd approximately)
	1 km = 0.621 mile ($\frac{5}{8}$ mile approximately)
Area	1 ft^2 = 0.093 m^2
	1 yd^2 = 0.836 m^2
	1 acre = 0.405 ha [1 ha (hectare) = 10 000 m^2]
	1 $mile^2$ = 2.590 km^2
	1 m^2 = 10.764 ft^2 = 1.196 yd^2 (approximately 1.2 yd^2)
	1 ha = 2.471 acres (approximately 2 $\frac{1}{2}$ acres)
	1 km^2 = 0.386 $mile^2$
Volume	1 ft^3 = 0.028 m^3
	1 yd^3 = 0.765 m^3
	1 m^3 = 35.315 ft^3 = 1.308 yd^3 (approximately 1.3 yd^3)
	1 ft^3 = 28.32 litres (1000 litres = 1 m^3)
	1 gal = 4.546 litres
	1 litre = 0.220 gal (approximately 4 $\frac{1}{2}$ litres to the gallon)
Mass	1 lb = 0.454 kg (kilogram)
	1 cwt = 50.80 kg (approximately 50 kg)
	1 ton = 1.016 t (1 tonne = 1000 kg = 0.984 ton)
	1 kg = 2.205 lb (approximately 2 $\frac{1}{5}$ lb)
Density	1 lb/ft^3 = 16.019 kg/m^3
	1 kg/m^3 = 0.062 lb/ft^3
Velocity	1 ft/s = 0.305 m/s
	1 mile/h = 1.609 km/h

Energy 1 therm = 105.506 MJ
1 Btu = 1.055 kJ

Thermal conductivity 1 Btu/ft^{2}h°F = 5.678 W/m^2°C

Temperature
x°F = $\frac{5}{9}$ (x − 32)°C
x°C = $\frac{9}{5}$x + 32°F
0°C = 32°F (freezing)
5°C = 41°F (cold)
10°C = 50°F (rather cold)
15°C = 59°F (fairly warm)
20°C = 68°F (warm)
25°C = 77°F (hot)
30°C = 86°F (very hot)

Pressure
1 lbf/in.^{2} = 0.0069 N/mm^2 = 6894.8 N/m^2
(1 MN/m^2 = 1 N/mm^2)
1 lbf/ft^2 = 47.88 N/m^2
1 tonf/in.^{2} = 15.44 MN/m^2
1 tonf/ft^2 = 107.3 kN/m^2

For speedy but approximate conversions

$$1 \text{ lbf/ft}^2 = \frac{\text{kN/m}^2}{20}, \text{ hence } 40 \text{ lbf/ft}^2 = 2 \text{ kN/m}^2$$

$$\text{and tonf/ft}^2 = \text{kN/m}^2 \times 10, \text{ hence } 2 \text{ tonf/ft}^2 = 20 \text{ kN/m}^2$$

Floor loadings
office floors – general usage: 50 lbf/ft^2 = 2.50 kN/m^2
office floors – data-processing equipment: 70 lbf/ft^2 = 3.50 kN/m^2
factory floors: 100 lbf/ft^2 = 5.00 kN/m^2

Safe bearing capacity of soil
1 tonf/ ft^2 = 107.25 kN/m^2
2 tonf/ft^2 = 214.50 kN/m^2
4 tonf/ft^2 = 429.00 kN/m^2

Stresses in concrete
100 lbf/in.^{2} = 0.70 MN/m^2
1000 lbf/in^2 = 7.00 MN/m^2
3000 lbf/in.^{2} = 21.00 MN/m^2
6000 lbf/in.^{2} = 41.00 MN/m^2

Costs
£1/m^2 = £0.092/ft^2
1 shilling (5p)/ft^2 = £0.538/m^2

£1/ft^2 = £10.764/m^2 (approximately £11/m^2)
£5/ft^2 = £54/m^2
£10/ft^2 = £108/m^2
£20/ft^2 = £216/m^2
£30/ft^2 = £323/m^2

APPENDIX IV MANHOLE SCHEDULE

MH nr	Size Internally	Wall Thick-ness	Cover	Invert Level	Ground Level	Cover Level	Depth of Excav'n	Depth of MH to Invert	Main Channel		Branch Channels		Notes
									Size	One Type	One Side	Other Side	
1	685 × 460	½B	457 × 457 grade C table 6	25.320	25.850	25.900	0.680	0.580	100	S	1/100	1/100	Concrete bases 150 thick. Precast r.c. cover slabs 150 thick.
2	798 × 685	1B	ditto	24.440	25.510	25.600	1.220	1.160	100	C	2/100	–	All covers in cast iron to BS 497 Pt.1.
3	900 × 685	1B	610 × 457 grade B table 5	23.500	25.120	25.200	1.770	1.700	100—150	ST	1/100	2/100	*channels* S = straight C = curved T = taper Step irons to mhs over 750 deep at 300 ccs.

Bibliography

Royal Institution of Chartered Surveyors and Building Employers Confederation. *Standard Method of Measurement of Building Works: Seventh Edition (SMM7)* (1988)

Royal Institution of Chartered Surveyors and Building Employers Confederation. *Code of Practice for Measurement of Building Works* (1988)

Building Project Information Committee. *Common Arrangement of Work Sections for Building Works* (1987)

Building Project Information Committee. *Project Specification: A Code of Procedure for Building Works* (1987)

Building Project Information Committee. *Production Drawings: A Code of Procedure for Building Works* (1987)

Co-ordinating Committee for Project Information. *Co-ordinated Project Information for Building Works,* a guide with examples (1987)

Property Services Agency, Royal Institution of Chartered Surveyors and Building Employers Confederation. *SMM7 Library of Standard Descriptions* (1988)

Fletcher, Moore, Monk and Dunstone. *Shorter Bills of Quantities: The Concise Standard Phraseology and Library of Descriptions..* Builder Group (1986)

I. H. Seeley. *Advanced Building Measurement.* Macmillan (1989)

I. H. Seeley. *Building Technology.* Macmillan (1993)

I. H. Seeley. *Quantity Surveying Practice.* Macmillan (1984)

I. H. Seeley. *Building Economics.* Macmillan (1983)

Society of Chief Quantity Surveyors in Local Government. *Life Cycle Cost Planning* (1984)

Royal Institution of Chartered Surveyors. *Guide to Life Cycle Costing for Construction* (1986)

Royal Institution of Chartered Surveyors. *Life Cycle Costing: A Worked Example* (1987)

W. James. Why the Quantity Surveyor? *Chartered Surveyor* (1960) pp. 602–6

R. C. Smith. *Estimating and Tendering for Building Work.* Longman (1986)

Chartered Institute of Building. *Code of Estimating Practice* (1984)

Institution of Civil Engineers and Federation of Civil Engineering Contractors. *Civil Engineering Standard Method of Measurement: Third Edition.* Telford (1991)

I. H. Seeley. *Civil Engineering Quantities.* Macmillan (1993)
Joint Contracts Tribunal. *Standard Form of Building Contract with Quantities* (1980)
The Aqua Group. *Pre-contract Practice for the Building Team.* BSP (1992)
The Aqua Group. *Tenders and Contracts for Building.* BSP (1990)
Society of Chief Quantity Surveyors in Local Government. *The Presentation and Format of Standard Preliminaries for use with JCT Form of Building Contract with Quantities 1980 Edition* (1981)
Greater London Council. *Preambles to Bills of Quantities.* Architectural Press (1980)
Royal Institution of Chartered Surveyors. *Definition of Prime Cost of Daywork carried out under a Building Contract* (1981)
Royal Institution of Chartered Surveyors. *Schedule of Basic Plant Charges for use in connection with Dayworks under a Building Contract* (1990)
Royal Institution of Chartered Surveyors. *Quantity Surveying 2000: the Future Role of the Chartered Quantity Surveyor* (1991)

Index